H. S. Fogle

Developments in Petroleum Science, 22

microbial enhanced oil recovery

DEVELOPMENTS IN PETROLEUM SCIENCE

Advisory Editor: G.V. Chilingarian

1. A.G. COLLINS
GEOCHEMISTRY OF OILFIELD WATERS

2. W.H. FERTL
ABNORMAL FORMATION PRESSURES

3. A.P. SZILAS
PRODUCTION AND TRANSPORT OF OIL AND GAS

4. C.E.B. CONYBEARE
GEOMORPHOLOGY OF OIL AND GAS FIELDS IN SANDSTONE BODIES

5. T.F. YEN and G.V. CHILINGARIAN (Editors)
OIL SHALE

6. D.W. PEACEMAN
FUNDAMENTALS OF NUMERICAL RESERVOIR SIMULATION

7. G.V. CHILINGARIAN and T.F. YEN (Editors)
BITUMENS, ASPHALTS AND TAR SANDS

8. L.P. DAKE
FUNDAMENTALS OF RESERVOIR ENGINEERING

9. K. MAGARA
COMPACTION AND FLUID MIGRATION

10. M.T. SILVIA and E.A. ROBINSON
DECONVOLUTION OF GEOPHYSICAL TIME SERIES IN THE EXPLORATION FOR OIL AND NATURAL GAS

11. G.V. CHILINGARIAN and P. VORABUTR
DRILLING AND DRILLING FLUIDS

12. T.D. VAN GOLF-RACHT
FUNDAMENTALS OF FRACTURED RESERVOIR ENGINEERING

13. J. FAYERS (Editor)
ENHANCED OIL RECOVERY

14. G. MÓZES (Editor)
PARAFFIN PRODUCTS

15A. O. SERRA
FUNDAMENTALS OF WELL-LOG INTERPRETATION, 1
THE ACQUISITION OF LOGGING DATA

15B. O. SERRA
FUNDAMENTALS OF WELL-LOG INTERPRETATION, 2
THE INTERPRETATION OF LOGGGING DATA

16. R.E. CHAPMAN
PETROLEUM GEOLOGY

17A. E.C. DONALDSON, G.V. CHILINGARIAN and T.F. YEN (Editors)
ENHANCED OIL RECOVERY, I
FUNDAMENTALS AND ANALYSES

18A. A.P. SZILAS
PRODUCTION AND TRANSPORT OF OIL AND GAS, A
FLOW MECHANICS AND PRODUCTION (second completely revised edition)

19A. G.V. CHILINGARIAN, J.O. ROBERTSON Jr. and S. KUMAR
SURFACE OPERATIONS IN PETROLEUM PRODUCTION, I

19B. G.V. CHILINGARIAN, J.O. ROBERTSON Jr. and S. KUMAR
SURFACE OPERATIONS IN PETROLEUM PRODUCTION, II

20. A.J. DIKKERS
GEOLOGY IN PETROLEUM PRODUCTION

21. F. RAMIREZ
APPLICATION OF OPTIMAL CONTROL THEORY TO ENHANCED OIL RECOVERY

Developments in Petroleum Science, 22

microbial enhanced oil recovery

Edited by

ERLE C. DONALDSON

School of Petroleum and Geological Engineering, University of Oklahoma, Norman, Oklahoma, U.S.A.

GEORGE V. CHILINGARIAN

Department of Petroleum Engineering, University of Southern California, Los Angeles, California, U.S.A.

and

TEH FU YEN

Departments of Civil and Environmental Engineering, University of Southern California, Los Angeles, California, U.S.A.

ELSEVIER – Amsterdam – Oxford – New York – Tokyo 1989

ELSEVIER SCIENCE PUBLISHERS B.V.
Sara Burgerhartstraat 25
P.O. Box 211, 1000 AE Amsterdam, The Netherlands

Distributors for the United States and Canada:

ELSEVIER SCIENCE PUBLISHING COMPANY INC.
655, Avenue of the Americas
New York, NY 10010, U.S.A.

LIBRARY OF CONGRESS
Library of Congress Cataloging-in-Publication Data

Microbial enhanced oil recovery / edited by Erle C. Donaldson, George V. Chilingarian, Teh Fu Yen.
p. cm. -- (Developments in petroleum science ; 22)
Includes bibliographies and indexes.
ISBN 0-444-42866-6 : fl 180.00 (est.)
1. Secondary recovery of oil. 2. Petroleum--Microbiology.
I. Donaldson, Erle C. II. Chilingarian, George V., 1929-
III. Yen, Teh Fu, 1927- . IV. Series.
TN871.M446 1988
665.5'3--dc19 87-27777
CIP

ISBN 0-444-42866-6 (Vol. 22)
ISBN 0-444-41625-0 (Series)

Printed in The Netherlands

Dedication

This book is dedicated to microbiologists whose innovative research helped to guide the direction of research of this newly developing aspect of enhanced oil recovery:

Dr. Dan Bubela, Australia
Dr. Edward Grula, U.S.A.
Dr. James Zajic, U.S.A.
Dr. S.I. Kuznetsov, U.S.S.R.
Dr. Claude E. ZoBell, U.S.A.

and to

Armand Hammer, Fred L. Hartley and Robert Wycoff

for their important contributions to the petroleum industry.

PREFACE

Crude oil is the major basic energy mineral for most industrial countries. It is apparent that there is no economical abundant substitute for crude oil. In much of the world additional supplies of crude oil will be developed by exploration and development of as yet undiscovered reserves. In the United States few such opportunities exist. Ultimately this will be true for all producing areas. For such areas, crude oil production will be sustained only by applying enhanced oil recovery (EOR) technologies. These EOR technologies are complex and resource exploitation using them will require more skilled personnel and substantial capital resources. Many of the EOR technologies are at advanced states of development, yet they have not yet been widely applied.

One such technique is microbial enhanced oil recovery (MEOR). The use of microorganisms to increase the recovery of oil from petroleum reservoirs was suggested in 1926 by Beckman and fundamental experiments and patents were developed by ZoBell by 1947. In spite of promising results in the decades of the 50's and 60's, these methods were not further developed or widely tested. Renewed interest in microbial methods began 10 years ago and has resulted in many field applications with mixed results because the biological, chemical and physical processes that occur in petroleum reservoirs where in-situ metabolism occurs are not well understood and the reasons for success could not be identified and easily repeated. As importantly, the reasons for failure could not be isolated and thus avoided in subsequent field applications.

This book reviews the substantial body of scientific and technical understanding that has been developed about the phenomenological processes. It does not pretend that the work is finished but organizes and summarizes work that has been done in by grouping it in important topical areas. The authors are actively involved in the study of microbes and microbial processes that can have impacts on production, transportation and storage of petroleum. Consequently, the book is more than a review of the literature, it contains the latest developments in MEOR.

Three of the authors have died since their contributions to this monograph were submitted. B. Bubela, E.A. Grula and J.E. Zajic were MEOR pioneers and I am pleased that the editors have dedicated this volume to their memory. Their contributions and enthusiasm did much to rejuvenate and stimulate interest in this field.

R.M. KNAPP
University of Oklahoma

LIST OF CONTRIBUTORS

T. BAN	*Faculty of Engineering, Shizuoka University, Hamamatsu, Shizuoka, Japan*
D. BRYANT	*Department of Food Technology, University of Illinois, Urbana, Ill., U.S.A.*
B. BUBELA[†]	*Baas Becking Geobiological Laboratory, P.O. Box 378, Canberra, A.C.T., Australia*
G.V. CHILINGARIAN	*School of Engineering, University of Southern California, University Park, Los Angeles, Calif., U.S.A.*
J.W. COSTERTON	*Department of Microbiology, University of Calgary, Calgary, Alta., Canada*
E.C. DONALDSON	*College of Engineering, School of Petroleum and Geological Engineering, University of Oklahoma, Norman, Okla., U.S.A.*
E.A. GRULA[†]	*Department of Microbiology, Oklahoma State University, Stillwater, Okla., U.S.A.*
M.M. GRULA	*Department of Microbiology, Oklahoma State University, Stillwater, Okla., U.S.A.*
A. GURROLA	*Department of Biology, The University of Texas at El Paso, El Paso, Texas, U.S.A.*
T.R. JACK	*NOVA Husky Research Corporation, 1411 25th Avenue, N.W., Calgary, Alta., Canada*
L.K. JANG	*Department of Chemical Engineering, California State University, Long Beach, Calif., U.S.A.*
G.E. JENNEMAN	*Department of Botany and Microbiology, University of Oklahoma, Norman, Okla., U.S.A.*
M. KENAGA	*Department of Food Technology, University of Illinois, Urbana, Ill., U.S.A.*
R.M. KNAPP	*College of Engineering, School of Petroleum and Geological Engineering, University of Oklahoma, Norman, Okla., U.S.A.*
R. PENNINGTON	*Department of Microbiology, Oklahoma State University, Stillwater, Okla., U.S.A.*
H.H. RUSSELL	*Department of Microbiology, Oklahoma State University, Stillwater, Okla., U.S.A.*

W. SEFFENS	*Department of Biology, The University of Texas at El Paso, El Paso, Texas, U.S.A.*
G.W. SEWELL	*Department of Microbiology, Oklahoma State University, Stillwater, Okla., U.S.A.*
J. SHAW	*Department of Microbiology, University of Calgary, Calgary, Alta., Canada*
N. WARDLAW	*Department of Chemistry, University of Calgary, Calgary, Alta., Canada*
T.F. YEN	*Departments of Civil and Environmental Engineering, University of Southern California, Los Angeles, Calif., U.S.A.*
J.E. ZAJIC[†]	*Department of Biology, The University of Texas at El Paso, El Paso, Texas, U.S.A.*

CONTENTS

Chapter 1

INTRODUCTION

ERLE C. DONALDSON, GEORGE V. CHILINGARIAN and TEH FU YEN

THE NEED FOR MICROBIAL ENHANCED OIL RECOVERY

The technology used today for the production of petroleum from subsurface reservoirs has not advanced beyond the stage at which the ultimate production of oil is only one third to one half of the original oil-in-place. Thus the potential target for enhanced oil recovery is greater than the reserves that can be produced by conventional methods (Yen, 1986). Efforts to improve the oil recovery efficiency through chemistry are marginally successful in making a small economical increment in oil recovery. Clearly another approach to enhancement of oil recovery is indicated. A new approach was initiated in 1980 by the U.S. Department of Energy by sponsorship of several research programs at universities to test the potential for enhancement of oil recovery using microbes. The suggestion that bacteria could be used for oil recovery was made by Beckmann (1926), and ZoBell (1947) who conducted additional experiments that indicated the potential for microbial release of oil from sand grains. Soviet scientists, Kuznetsov (1961), Kuznetsov et al. (1962), Ivanov et al. (1982), and Belyaev (1983), also made important contributions to the field of microbial enhancement of oil recovery.

The results of the U.S. Department of Energy sponsored research, and other work around the world, were published in the proceedings of two international conferences that were held on microbial enhancement of oil recovery (MEOR) (Donaldson and Clark, 1983; Zajic and Donaldson, 1985).

Microbial cultures are capable of synthesizing a large variety of biochemical products from crude oil constituents when provided with essential nutrients and proper environmental conditions. In fact, populations of various types of microbes have been found in subsurface petroleum reservoirs up to 2000 m in depth (Petzel and Williams, 1986). The prevalent theory is that the microbes were introduced by drilling, well completion, well workover procedures, or by water injection during secondary recovery. There is a small amount of work in progress at the University of Oklahoma to determine if microbial spores can survive through geological time in subsurface reservoirs. Whether recently introduced or germinated by pressure–temperature changes caused by production of oil and water, the fact that the populations of microbes are present in deep porous formations is positive proof that petroleum reservoirs are not biologically sterile and offers positive encouragement for some very fundamental research to discover microbes, procedures, and technology that can lead to enhancement of oil recovery.

The range of metabolic products from microbial attack of petroleum is very broad, depending on environmental conditions (pressure, temperature, salinity, pH, and the presence or absence of oxygen), supporting nutrients available for cell metabolism (nitrogen, phosphorus, etc.) and the specific bacteria interacting with the petroleum. In very general terms, the metabolic products may be gases (methane, hydrogen, carbon dioxide, hydrogen sulfide), carboxylic acids (formic, acetic, valeric), solvents (alcohols, aldehydes, ketones), polymers (proteins, polysaccharides), surface-active compounds (poly-anionic lipids) and many other compounds ranging from simple to very complex macromolecules. The chemical structure of many of the high-molecular-weight polymers produced by microbes have not been elucidated because of their molecular diversity and complexity. The renewed emphasis on these products for possible application to petroleum recovery and processing, however, should result in greater emphasis on the determination of the chemical structure and properties of microbial bioproducts.

With respect to petroleum, the study of microbes and their products is carried out to: (1) develop products that can enhance secondary and tertiary recovery; (2) cause the mobilization of heavy oils by viscosity and interfacial tension reduction; (3) enable injection of cells into petroleum reservoirs, which will produce bioproducts in situ to enhance the recovery of oil; and (4) study of petroleum reservoir microbial ecology. The emphasis of this book is the first of these with the objectives of showing some of the developing technologies and the potential that exists for microbial enhancement of oil recovery.

The practical application of microbial cultures to subsurface oil reservoirs imposes several restrictions on the microbial cultures. The microbes must be able to migrate, or to be transported, deep within the reservoir for any in-situ applications to be of practical significance to oil recovery. Furthermore, they must be able to multiply in the subsurface environment and, therefore, nutrients required for growth which are not available in a petroleum reservoir must be supplied in the injection water.

Microbial toxicity to heavy metals increases at elevated temperature; and, since heavy metals are frequently present in reservoir brines, the increase of temperature in the subsurface (25°C + 18°C/km of depth) might inhibit the effective in situ application of microbial cultures. Consequently, microbes intended for use in petroleum reservoirs should be tested with the reservoir fluids at subsurface conditions of temperature and pressure. Bubela (1983) examined the effects of reservoir environmental conditions on microbial cultures and observed increased toxicity to copper, and changes of morphology from rod-shaped to spherical, coccoidal, forms when subjected to 60°C and 20,000 kPa pressure. If a change of metabolism (involving nutrient requirements and metabolic products) accompanies the change of morphology, microbes that are apparently useful for EOR when tested at ambient conditions may become useless under reservoir conditions. These aspects of microbial EOR have not been thoroughly tested, but they may explain some of the negative results that have been reported for direct field applications. To support this thesis, Marquis (1983) reported that high pressures reduce the growth rate of microbes at all temperatures. Microbial systems intended for subsurface applications, therefore, should be tested at simulated subsurface environmental conditions, or selected microbial cultures should be isolated under these conditions. Perhaps a better source for the search for microorganisms that can enhance oil recovery in situ would be the petroleum reservoir itself. The ecology of petroleum reservoirs has not been thoroughly explored, yet populations of aerobic bacteria are known to grow around injection wells, and facultative and anaerobic bacteria have been isolated from deep reservoirs. Grula and Sewell (1983) isolated a facultative bacterium along with anaerobic *Desulfovibrio* from crude oil and brine samples from a depth of 1000 m in the Wilmington oilfield, Long Beach, California. The facultative organism is a Gram-negative rod, motile, non-spore forming, and produces a viscous biopolymer. It showed a peculiar synergistic relationship to *Desulfovibrio* when grown together. *Desulfovibrio* exhibited more rapid growth than usual and enhanced production of hydrogen sulfide. The implications of this discovery to petroleum production are not good because increased production of hydrogen sulfide may result in additional corrosion and the viscous biopolymer may afford protection against normal biocides used to control *Desulfovibrio*. Although these results were not promising, as far as MEOR is concerned, the discovery of a viable facultative, polymer-producing, bacterium deep in an oilfield is significant, because it implies that other bacteria may reside in petroleum reservoirs that may indeed have an impact on the mobilization of residual oil.

Lazar (1983) isolated mixed cultures of bacteria from various locations that were then adapted to reservoir conditions. Species identified in mixed cultures were: *Pseudomonas, Escherichia, Arthrobacter, Mycobacterium, Micrococcus, Peptococcus, Bacillus*, and *Clostridium*. They found that the mixed cultures were more efficient in releasing oil than pure strains. Seven oilfields were inoculated, but only two of the oilfields responded favorably with increased oil production ranging from 16 to 200%. The increased oil production in the two fields, however, continued for two years. The failure of the other oilfields to respond favorably was attributed to high salinity and low permeability; hence, as is so often the case in fields designated for research, the fields were very poor choices for definitive pilot studies of MEOR.

The idea of ex-situ production of a surfactant-type bioproducts from oilfield crude oil treated with mixed cultures of aerobic bacteria for subsequent addition to injection water is under investigation in China by Zhang and Qin (1983). They have found that aerobic fermentation of crude oil yields a biosurfactant that readily forms a microemulsion with crude oil and water, which has a greatly reduced viscosity compared to the crude oil. When the fermentation products are mixed with crude oil having a viscosity of 2500 centistokes, in a 1 : 1 ratio, the viscosity of the resulting mixture is between 12 and 46 centistokes. If the fermentation can be carried out economically on a large scale and applied to an oilfiled, especially one with a high saturation such as the viscous oil that was mentioned, the possibility of an enormous increase of oil production certainly exists. The biosurfactants exhibit an advantage over most synthetic detergents because they are not as sensitive to dissolved salts which are present in most oilfield waters. Hitzman (1983) reviewed the literature on field applications of microbial enhancement of oil recovery. More than 200 field tests were conducted. Although in most cases the wells that were inoculated were located in oilfields of very poor quality (heterogeneous, low porosity and permeability, and low oil saturation), many responded favorably to microbial treatment. As experience was gained in field applications, the techniques became more complex and technically advanced. The researchers began with simple inoculations of single wells without pretreatment and advanced to more complex injections of low-salinity water buffer zones, injection of various nutrients, establishment of a fermentation period for cell growth, etc. In some cases, the procedures were simply single-well clean-outs, where bacteria such as *Clostridium* are used to build up a high pressure around the production well which is then suddenly released to displace fines and other debris into the wellbore. In other cases, the objective was long-term stimulation of oil recovery from a large portion of the reservoir. In-situ microbial growth apparently resulted in chemical and petrophysical changes within the reservoir that led to positive microbial enhancement of oil recovery. A significant conclusion from the review was that the reservoir temperature and pressure may not be as restrictive to MEOR as indicated by specific laboratory conditions. This question, however, remains to be resolved by carefully conducted laboratory and field experiments.

The few oilfield successes that have been reported imply that MEOR is potentially economical and feasible, but the failures remain to be completely understood. Specific research for field application should resolve some of these in the future.

SYNOPSIS OF THE BOOK

Chapter 2 begins with a brief review of the subsurface environment as a system for growth and transport of microorganisms. Subsurface porous and permeable geologic formations present an environment which is very different from the usual laboratory environments used for the growth of microorganisms. The temperature increases with depth, from mean surface temperature, by an average of 18°C/km of depth. Thus at a depth of 1000 m the temperature will be approximately 45°C, which is the limiting growth temperature for many species of bacteria under laboratory conditions. Some microbes, however, can survive in petroleum reservoirs up to a depth of 2000 m (Petzel and Williams, 1986), where temperature and fluid pressure are about 63°C and 20 MPa, respectively. The morphology of the microbes was changed from rod-shaped to spherical, probably due to the influence of the high pore pressure. These microbes apparently have adapted to the high salinity of the reservoir, which is another inhibitory influence to bacterial growth in the laboratory.

They also apparently remain as viable cells for very long periods of time in the absence of rich mixtures of nutrients used under laboratory conditions for microbial growth. Thus, the subsurface environment analyzed superficially for its effects on the growth of microbes is apparently inhibitive toward growth. The high temperature and the high pressure may be limiting to growth; the high salinity is far beyond the limits for microbial growth at laboratory conditions; essential nutrients (phosphates, oxygen, nitrates, etc.) are absent; and the porous medium inhibits transportation of the microbes. The subsurface environment must be capable of sustaining microbial growth and, therefore, any microbial cultures proposed for in-situ applications should be studied in equipment that can simulate the actual subsurface environment as a whole. The microbes should be placed in rocks containing brine, at a pore pressure and temperature comparable to those at some specific subsurface depth (Donaldson et al., 1980).

Oil reservoirs that have reached the limit of economic oil production by secondary recovery will probably be those that are used for MEOR whenever cultures suitable for MEOR at subsurface conditions have been isolated. Microbes that follow the preexisting fluid flow paths from injection wells to producing wells probably will not prove beneficial unless the microbes establish blocks to fluid flow in the larger pores by producing biopolymers, and other metabolic products or biomass from cell growth and adsorption to the pore surfaces (discussed in greater detail in Chapter 3). This would improve the sweep efficiency of the injected water, and also divert the water into zones where high oil saturations were bypassed, such as low-permeability zones. If the injected microbes produce large quantities of biopolymers, some improved recovery would probably result by improved mobility ratio of the injected fluid. On the other hand, some microbes produce copious amounts of carbon dioxide in an anaerobic environment, such as a petroleum reservoir; therefore, these bacteria may be used for increasing the overall reservoir pressure if their growth is controlled for a long period of time or for release of oil in situ by generation of carbon dioxide at the oil–water–rock interfaces.

Microbes in an anaerobic environment apparently do not consume the hydrocarbon constituents of petroleum and, thus, petroleum does not act as a nutrient for the microbes. Thus, nutrients (dissolved in injected waters) required by anaerobic bacteria must be supplied in sufficient quantities to sustain the growth of microbes in situ. Experiments in oilfields (Donaldson and Clark, 1983) indicate that periodic inoculations of the microbes along with continuous injection of a dilute solution of essential nutrients, is required for long-term (1 year or more) sustained growth of microbes in a petroleum reservoir. Microbes such as *Desulfovibrio* exist in reservoirs for long periods of time and the growth of introduced cultures for MEOR have been maintained for long periods of time by periodic inoculations of the reservoir.

Chapter 3 constitutes a specific discussion of the potential for in-situ microbial applications for MEOR. The oil left in place is trapped by capillary forces and chemical adhesion to the reservoir minerals. Thus, efforts to displace this residual oil require addition of considerable energy to the petroleum reservoir in the form of chemical reactants, heat, or solvents. A potential source of supplemental energy for petroleum reservoirs and other petroleum processing operations is from microorganisms. Microbes of various types (aerobic, anaerobic, mesophilic, and the thermophilic) produce a large number of chemical compounds as products of their metabolism. The type of product and quantitative yield depends on the specific microbial colony, nutrients provided for its growth, and prevailing environmental conditions. Within the range of products are: the gases—carbon dioxide, hydrogen, and methane; polymers, which are either polysaccharides or proteins; surface-active compounds that are generally poly-anionic lipids; low-molecular-weight solvents, such as alcohols and ketones; and low-molecular-weight carboxylic acids. Laboratory experiments have shown that (1) some of the metabolites are capable of releasing oil from packed sand columns and rocks, (2) some cultures can reduce the viscosity of heavy oils, and (3) the polymers are useful for plugging high-permeability zones and displacing oil through mobility control. Precisely controlled pilot experiments under joint supervision of microbiologists and engineers are necessary to scale the processes to field quantities for development of this potentially new technology. The type and yield of metabolic products can be controlled to a large extent by modifications of environmental conditions and nutrients. Additionally, the microbes exhibit amazing properties for adaptation to new circumstances, both environmentally and nutritionally, through mutation. These properties give the engineer considerable flexibility for applications of microbial systems (Grist, 1983).

This leads to a discussion of geobiology in Chapter 4. When microbial activity develops in a subsurface geological environment, the geological, mineralogical, hydrological, and geochemical aspects of the environment will have a profound effect on the microorganisms and, in turn, the microbial population will have some effect on the rocks and fluids. The most significant geological changes are: (1) precipitation of dissolved minerals, especially carbonates; (2) change of permeability due to precipitation in pore throats; and (3) a change of porosity, either increase or decrease, depending on the equilibria of dissolved salts and products of organic acids.

Following the generally theoretical discussions of the interactions between microbes and petroleum reservoirs, Chapter 5 leads directly into a discussion of laboratory observations of oil recovery from Hele-Shaw models. The Hele-Shaw model consists of a viscous fluid confined between closely spaced pieces of glass. When the viscous fluid is displaced by a less viscous fluid, frontal instability (fingering), sweep efficiency, and interfacial effects can be closely observed and photographed. Using these models, strains of *Zymomonas* were studied for oil displacement efficiency, and oil displacement by biologically produced polysaccharides was examined. Then follows Chapter 6 on laboratory and field applications of anaerobic and facultatively anaerobic bacteria for oil displacement. Living cells were found to adsorb on sandstone, suggesting that spores of such bacteria as *Clostridium* may be a better means of developing a large microbial population deep within a reservoir. Salt-tolerant *Clostridium* was shown to increase oil production by in-situ production of gas and solvents.

A discussion of reduction of permeability by aerobic bacteria cells around injection wells to deep plugging of high-permeability zones within the reservoir is presented in Chapter 7. The discussion encompasses the geologic effects of bacteria that have so far been discovered in petroleum reservoirs. Thus, this chapter explores microbial plugging problems that have been reported in the literature and evaluates the mechanisms responsible for the plugging effects. The characteristics of microbial cultures which are required for successful injection of microbes for in situ MEOR are examined. After considering that bacteria can cause reduced injectivity by growing in the vicinity of injection wells, the problem is approached in Chapter 8 from the point of view of the transport mechanisms of bacteria in porous media. The migration and growth of cells through nutrient-enriched, oil-containing, reservoir rocks is discussed from direct experience with laboratory samples and review of the current literature on the topic of microbial transport in porous media.

Chapter 9 is a short chapter considering the optimum hydrophobic–hydrophilic balance for growth of microbes and production of biosurfactants. The authors approach this topic from an experimental basis and show that the hydrophobic–hydrophilic balance of the growth medium is important in hydrocarbon fermentations.

Chapter 10 examines the problem of biodegradation of chemical compounds used in enhanced oil recovery operations. This presents a costly problem in the use of xanthan gum as a mobility control agent, because aerobic bacteria attack it as soon as it is dissolved in water. Thus, the solutions must be protected with biocides and maintained in closed, oxygen-free, tanks. Also polyacrylamides are readily hydrolyzed to polyacrylic acid and the ammonia can then be utilized by bacteria indigenous to a petroleum reservoir, such as *Desulfovibrio*, producing a bloom of growth and increasing hydrogen sulfide production. The factors affecting biodegradation and the rates of degradation are discussed as well as some measures of abatement.

The final chapter, Chapter 11, points out some of the potential health hazards of bacteria used in MEOR. This is an aspect of MEOR that has been largely ignored in the literature; therefore, this book concludes with a short, but poignant, discussion of health hazards.

TECHNICAL CONSTRAINTS OF ENHANCED OIL RECOVERY

Current EOR technology is developed to recover as much as possible of the residual oil after primary and secondary (or improved) waterflooding. Reviews of the literature (BETC Staff, 1980; Yen, 1982) reveal that all of the current EOR techniques suffer from many shortcomings and constraints. Some of the problems of the individual, traditional EOR methods are discussed briefly:

(1) *Miscible flooding*: The carbon dioxide miscible EOR process is ideal for deep reservoir, light oil operations. It has been shown that the ratio of CO_2 to oil is 8 Mcf/bbl for a profitable venture with 3 Mcf/bbl remaining in the reservoir. The drawback is early and premature breakthrough. Prediction of accurate miscibility pressure, single contact phase properties, and dynamic phase behavior is tentative or unknown. Asphaltenes also can be precipitated by CO_2. Asphaltenes can, however, be made soluble in the CO_2–water system if the Hildebrand solubility parameter criterion is met.

(2) *Surfactant flooding*: Present knowledge indicates adsorption, chemical reactions with various ions and clay minerals in the reservoir, and temperature instability are some of the most destructive factors in the case of surfactant flooding. Another important fact is that the surfactant may react with the crude oil components to render the surfactant inactive (Yen, 1982); that is, the commonly-used anionic petroleum sulfonate type of surfactant can react with an organic base from the crude (such as pyridine) to form water-insoluble salts and nullify the activity.

(3) *Polymer flooding*: Both synthetic polymers (such as polyacrylamides) and natural polymers (such as polysaccharides) are used for improvement of sweep efficiency. The main problem with polymer usage is its mechanical (shear), chemical, and microbiological stability. Also, reduction of permeability has been experienced in field trials (Smith and Fleming, 1980; Harpole and Hall, 1983). One of the problems is ionization occurring when polyamide is in contact with organic acids in the petroleum. Crosslinked networks will subsequently form macroscopic particles which can plug the pores.

(4) *Fireflooding*: Although in-situ combustion has the advantage of supplying its own fuel, gravity override as well as casing, tubing, pump and well damage are encountered. One of the ill effects is the conversion of premium petroleum into useless coke underground.

(5) *Steamflooding*: Use of steam in EOR has become a mature technology, yet problems still exist for high heat loss and low energy utilization. Adequate water supply is always a constraint. Furthermore, there is no good method for improving sweep-efficiency. Steam override amplified by steam distillation of petroleum in multiphase flow often reduces sweep efficiency. Usually, additives are required to stabilize the heat sensitive fractions as well as to cause foaming. The heat sensitive portion of petroleum, that is, the aromatized napthenics, can cause polycondensates originating from Diels-Alder type reactions (Yen, 1982). These polycondensates can cause plugging.

(6) *Alkaline flooding*: Aside from the ordinary problems, the uncertainties in acid number and chemical loss make the exact mechanism of alkaline flooding unclear. There also is a lack of serious chemical studies on the interaction between alkali and the various petroleum components.

The above discussion deals with all the known approaches to EOR. In almost every case, there are technological and/or economical constraints to application of the processes. Some of the problems can be partially solved. Currently, the petroleum industry is exerting an effort to modify the EOR technology for specific cases and thus use these existing techniques on a limited basis.

MICROBIOLOGICAL ENHANCED OIL PROCESSES

The use of microorganisms and their metabolic products to stimulate oil production is now receiving renewed interest worldwide. This technique involves the injection of selected microorganisms into the reservoir and the subsequent stimulation and transportation of their in-situ growth products in order that their presence will aid in further reduction of residual oil left in the reservoir after secondary recovery is exhausted. The MEOR is unlikely to replace conventional EOR methods, because MEOR itself has certain constraints. This unique process seems superior in many respects, however, because self-duplicating units, namely the bacteria cells, are injected into the reservoir and by their in-situ multiplication they magnify their beneficial effects. Some of the mechanisms proposed, by which these microbial agents could stimulate oil release are shown in Table 1-I. The CO_2 and/or CH_4 produced from the fermentation and neutralization of acid products by the

TABLE 1-I
Microbial products and their contribution to enhanced oil recovery (after Janshekar, 1985)

Bioproduct	Effect
Acids	Modification of reservoir rock
	Improvement of porosity and permeability
	Reaction with calcareous rocks and CO_2 production
Biomass	Selective or nonselective plugging
	Emulsification through adherence to hydrocarbons
	Modification of solid surfaces
	Degradation and alteration of oil
	Reduction of oil viscosity and oil pour point
	Desulfurization of oil
Gases (CO_2, CH_4, H_2)	Reservoir repressurization
	Oil swelling
	Viscosity reduction
	Increase of permeability due to solubilization of carbonate rocks by CO_2
Solvents	Dissolving of oil
Surface-active agents	Lowering of interfacial tension
	Emulsification
Polymers	Mobility control
	Selective or non-selective plugging

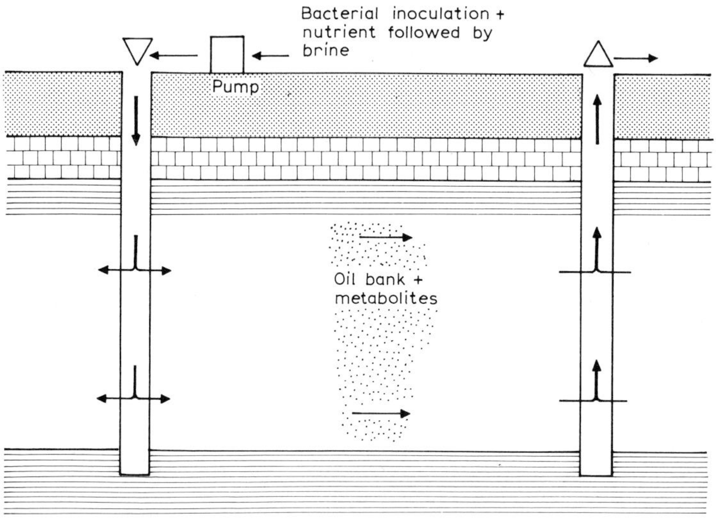

Fig. 1-1. Schematic showing displacement of oil by metabolites of inoculated bacteria grown in situ.

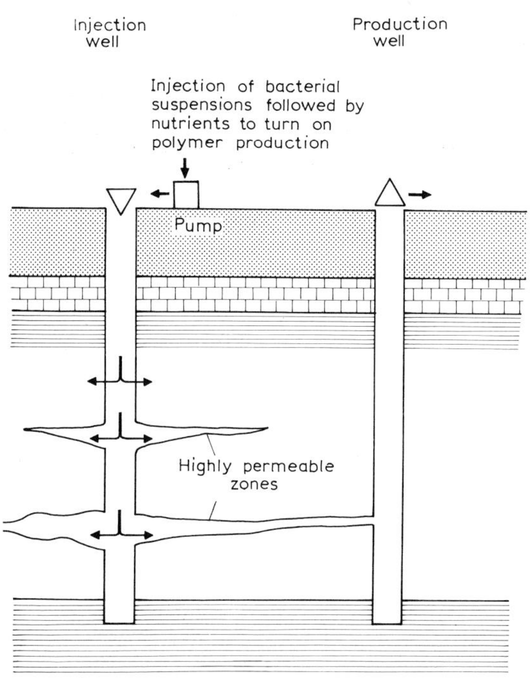

Fig. 1-2. Schematic showing selective plugging of highly permeable zones by bacteria. Production of extracellular polymers is "turned on" by injecting a nutrient solution containing sucrose.

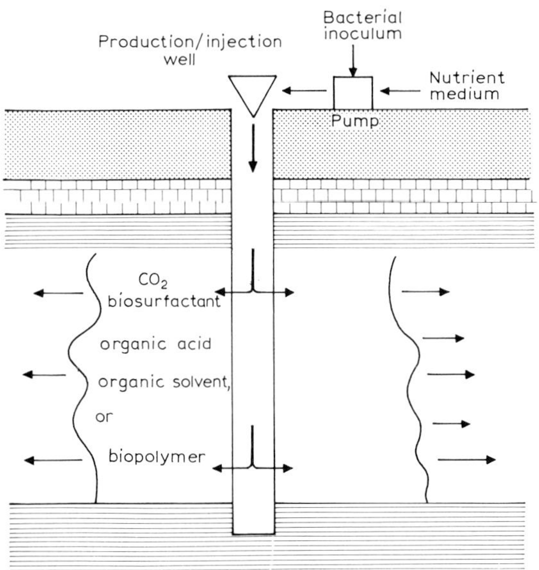

Fig. 1-3. Schematic showing the migration of cells and the synthesis of metabolic products around the wellbore following inoculation and closing of injection well. (This corresponds to the "huff" stage of "huff-and-puff" process.)

reservoir rock could lower the pH of the water and repressurize the reservoir. The solvents, primarily alcohols such as methanol, ethanol, propanol, isobutanol, *n*-butanol, as well as formaldehyde and acetone could reduce oil viscosity. Figs. 1-1 to 1-4 illustrate the methods of using bacteria for the recovery of residual oil.

Although production specialists have been concerned about the possible adverse effects of using microorganisms, such as fouling of the process water and souring of oil reservoirs with H_2S due to microbial action, it is now believed that such uncontrolled microbial action could be avoided by proper design of the process. This would lead to additional oil recovery after conventional recovery methods are exhausted.

From the early discoveries in petroleum microbiology, through the work of ZoBell (1947) and early field tests conducted in Eastern Europe, and to the present, the understanding of the nature of microbial enhancement of oil recovery has steadily increased. To date, however, many issues in MEOR are still debated and few systematic approaches are available for definitive evaluations. Reservoir mechanics and MEOR performance are complexly interrelated. It is clear from theoretical considerations and from experimental evidence that many factors could influence the penetration of cells in a porous matrix. These include: (1) physical and chemical properties of the rocks, such as permeability, pore size distribution, porosity, wettability, surface charge, type of oil (e.g. polar versus non-polar), and total salinity and ionic composition of formation water; (2) cell properties, such as shape,

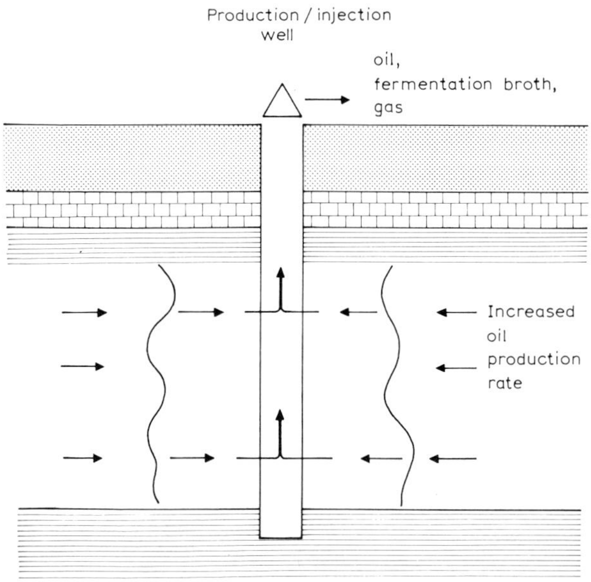

Fig. 1-4. Schematic showing the production of oil at the end of the incubation period, when the well is reopened. (This corresponds to the "puff" stage of "huff-and-puff" process.)

size, motility, type of cell growth (individually or in clumps or chains), surface charge, production of capsules and slimes, chemical reaction products (acids or gases); (3) mode of injection, such as rate of injection, salt content of injection water and density of cell suspension. After considering all of these factors, it is essential to treat the whole reservoir as a bioreactor and study the displacement efficiency and the applicability of releasing oil by bacteria under reservoir conditions. The wettability alteration and effective oil permeability modification when bacteria are injected into the reservoir have to be investigated. In any event, the addition of growth-promoting nutrients is required to obtain maximum bacterial activity of the type desired. Production of gas, acid, and surfactants throughout the oil-bearing formation is, theoretically, the principal objective. Ideally, without regard to development within the formation of extraneous microbes, cells or spores of the specific microbe to be employed may be introduced along with the nutrient, or before or after the nutrient is injected into the formation. It is only through multidisciplinary integration that MEOR can become a technically proven and feasible technology.

REFERENCES

Beckmann, J.W., 1926. The action of bacteria on mineral oil. *Ind. Eng. Chem. News*, 4 (Nov. 10): 3.

Belyaev, S.S., 1983. The dynamics of development of aerobic and anaerobic bacteria during aeration of an oil-bearing stratum to enhance oil recovery. In: T.F. Yen (Chairman), *Symposium on Biological*

Pressures Related to Petroleum Recovery. Div. Pet. Chem., Am. Chem. Soc., Seattle, Wash., March 20–25, pp. 810–812.

BETC Staff, 1980. *Technical Constraints Limiting Application of EOR Techniques to Petroleum Production in the United States* (revised September 1980, TI 80/4; January 1984, RI-83/9). NTIS, Springfield, Va.

Bubela, B., 1983. Combined effect of temperature and other environmental stresses on microbiologically enhanced oil recovery. In: E.C. Donaldson and J.B. Clark (Editors), *Proceedings, 1982 International Conference on Microbial Enhancement of Oil Recovery*. NTIS, Springfield, Va., pp. 118–123.

Donaldson, E.C. and Clark, J.B. (Editors), 1983. *Proceedings, 1982 International Conference on Microbial Enhancement of Oil Recovery*. NTIS, Springfield, Va., 219 pp.

Donaldson, E.C., Kendall, R.F., Pavelka, E.A. and Crocker, M.E., 1980. *Equipment and Procedures for Fluid Flow and Wettability Tests of Geologic Materials*. NTIS, Springfield, Va., DOE/BETC/IC-79/5, 40 pp.

Grist, D.M., 1983. Microbial enhancement of oil recovery—an operator's view. In: *Biotech 83. Proceedings International Conference on Commercial Applications of Biotechnology*. Online Publications, Northwood, pp. 463–474.

Grula, M.M. and Sewell, G.W., 1983. Microbial interactions with polyacrylamide polymers. In: E.D. Donaldson and J.B. Clark (Editors), *Proceedings, 1982 International Conference on Microbial Enhancement of Oil Recovery*. NTIS, Springfield, Va., pp. 129–134.

Harpole, K.H. and Hall, C.J., 1983. *Evaluation of the N. Stanley Demonstration Project*. NTIS, Springfield, Va., DOE/BC/10033-6, February.

Hitzman, D.O., 1983. Petroleum microbiology and the history of its role in EOR. In: E.C. Donaldson and J.B. Clark (Editors), *Proceedings, 1982 International Conference on Microbial Enhancement of Oil Recovery*. NTIS, Springfield, Va., pp. 162–218.

Ivanov, M.V., Belyaev, S.S., Laurinavichus, K.S., Obraztsova, A.Y. and Gorlatov, S.N., 1982. *Microbiologiya*, 4: 336.

Janshekar, H., 1985. Microbial enhanced oil recovery processes. In: J.E. Zajic and E.C. Donaldson (Editors), *Microbes and Oil Recovery, 1*. Bioresources Publications, El Paso, Texas, pp. 54–84.

Kuznetsov, S.I. (Editor), 1961. *Geologic Activity of Microorganisms* (translated from Russian by Consultants Bureau, New York, 1962). *Tr. Inst. Mikrobiol. Akad. Nauk SSSR*, IX, 112 pp.

Kuznetsov, S.I., Ivanov, M.V. and Lyalikova, N.N., 1962. *Introduction to Geological Microbiology* (translated from Russian by P.T. Broneer and C.H. Oppenheimer). *Introduction to Geological Microbiology*, McGraw-Hill, New York, N.Y., 252 pp.

Lazar, I., 1983. Microbial enhancement of oil recovery in Romania. In: E.C. Donaldson and J.B. Clark (Editors), *Proceedings, 1982 International Conference on Microbial Enhanced Oil Recovery*. NTIS, Springfield, Va., pp. 140–148.

Marquis, R.E., 1983. Barobiology of deep oil formations. In: E.C. Donaldson and J.B. Clark (Editors), *Proceedings, 1982 International Conference on Microbial Enhancement of Oil Recovery*. NTIS, Springfield, Va., pp. 124–128.

Petzel, G.A. and Williams, B., 1986. Operations trim basic EOR research. *Oil Gas J.*, 84/6 (Feb. 10): 41–45.

Smith, R.V. and Fleming, P.D., 1980. Fundamental mechanisms influencing recovery of crude oils. *Proc. 10th World Pet. Congr.*, 3: 127.

Yen, T.F., 1982. Chemical problems related to EOR. *Symp. on Chemical and Geochemical Aspects of Oil Recovery. Annu. Meet. Am. Chem. Soc., Kansas City, Sept. 1982.*

Yen, T.F., 1986. *A State-of-the-Art Review on Microbial Enhanced Oil Recovery*. Unpublished report, University of Southern California, Los Angeles, Calif., 278 pp.

Zajic, J.E. and Donaldson, E.C. (Editors), 1985. *Microbes and Oil Recovery 1*. Bioresources Publications, El Paso, Texas, 370 pp.

Zhang, Z. and Qin, T., 1983. A survey of research on the application of microbial techniques to the petroleum production in China. In: E.C. Donaldson and J.B. Clark (Editors), *Proceedings, 1982 International Conference on Microbial Enhancement of Oil Recovery*. NTIS, Springfield, Va., pp. 135–139.

ZoBell, C.E., 1947. Bacterial release of oil from oil-bearing materials, Parts I and II. *World Oil*, 126(13): 36–47 (I); 127(1): 35–41 (II).

Chapter 2

THE SUBSURFACE ENVIRONMENT

ERLE C. DONALDSON, ROY M. KNAPP, T.F. YEN and GEORGE V. CHILINGARIAN

INTRODUCTION

In considering the use of viable microbial cultures within petroleum reservoirs for enhancement of oil recovery (EOR) through metabolic activities of the microbes, one is immediately confronted with an unusual environment existing in the subsurface. Not only is this environment completely foreign to life systems adapted to surface conditions, but it changes in a generally predictable manner as the depth from the surface increases. The porosity and permeability of sedimentary formations decrease with depth of burial. The pressure and temperature increase with depth. The salinity of subsurface waters also commonly increases with depth.

Although certain types of microorganisms may grow prolifically at surface conditions and produce metabolic products that would obviously be beneficial to EOR, they may not be able to survive in a deep subsurface environment. If they do grow at subsurface conditions, they may not produce the same metabolic products because of the adverse influences of the high salt content and increased temperature of subsurface waters. In addition, microbes are known to undergo a change of morphology at high pressures which may cause a change of metabolism.

Thousands of different species of microorganisms have been classified and examined to determine their nutrient requirements, products of their metabolism, and the limits of environmental conditions that they can tolerate. Four very broad divisions have emerged based on environmental tolerance: (1) aerobes, capable of life processes only in the presence of oxygen; (2) anaerobes, which exist in the absence of oxygen and derive their energy from degradation of oxygenated molecules; (3) facultative, which are able to exist aerobically or anaerobically; and (4) a division of the three categories above as either mesophilic (capable of life below 45°C) or as thermophilic (able to live in environments at a constant temperature greater that 45°C). Within these broad groupings, there are microbes that metabolically produce gases (methane, hydrogen, nitrogen, carbon dioxide), polymers (polysaccharides, proteins), surface-active compounds that are generally poly-anionic lipids, and many other compounds from simple alcohols to very complex macromolecules. The type and yield of metabolic products can be controlled to a large extent by modifications of environmental conditions and nutrients. Additionally, the microbes exhibit amazing properties for adaptation to new circumstances, both environmentally and nutritionally, through mutation. These properties give the engineer considerable flexibility for applications of microbial systems (Donaldson, 1982).

Although many distinct types of microbes have been isolated and thoroughly characterized, this by no means implies that all microbes have been found. There may be many undiscovered microbial systems that are either already existing in deep subsurface reservoirs or are capable of existence there if the required nutrients are supplied for their metabolism. The potential for their discovery and use for oil recovery and other operations certainly exists.

SUBSURFACE BRINES

The amount of total dissolved solids (TDS) of subsurface interstitial waters commonly increases with depth, reaching values which approach supersaturation— more than 350 g/l at great depths. In outcropping sediments where meteoric water can circulate, the ground water may be comparatively fresh considering the depth of burial, and it will generally contain bicarbonate and sulfate anions. The deep underground waters, however, are stagnant and very saline with a concentration

TABLE 2-I

Average concentrations of ions in sea water and subsurface brines of various ages from sedimentary basins in the United States (from Collins, 1975, tables 5.II, 7.IV–7.XIII, with permission)

Constituent (mg/l)	Sea water	Tertiary	Cretaceous	Jurassic	Permian
Lithium	0.2	4	4	10	3
Sodium	11,000	39,000	31,000	57,300	47,000
Potassium	350	220	130	140	170
Rubidium	0.1	0.24	0.10	0.10	0.80
Cesium	–	0.20	0.10	0.10	0.13
Calcium	400	2,530	7,000	25,800	8,600
Magnesium	1,300	530	900	2,500	2,000
Strontium	7	130	200	320	7
Barium	–	60	40	10	–
Boron	5	36	20	13	8
Manganese	–	–	–	–	–
Copper	–	0.63	–	–	0.88
Chloride	19,000	64,600	62,000	141,000	92,700
Bromide	65	85	550	1,200	46
Iodide	0.05	28	25	16	3
Bicarbonate	–	560	260	140	77
Carbonate	–	75	–	–	36
Sulfate	–	320	280	210	730
Organic acid	–	140	–	12	170
Ammonium	–	230	23	–	24

gradient approaching 50,000 ppm per kilometer of depth, reaching a maximum of 350,000 ppm TDS (Donaldson, 1980).

Subsurface reservoirs containing hydrocarbons are generally sedimentary deposits consisting of sandstones and carbonate rocks originally deposited in a marine environment (Levorsen, 1966). The subsurface waters associated with hydrocarbons were in the rocks originally, or migrated into the porous subsurface rocks from the surface. The most generally accepted theory of the association of oil with ground waters is that the oil eventually migrated from the site of origin (the source rock) to porous sediments formed into porous traps by tectonic movements of the rocks, or by deposition of impermeable shales on top of a porous zone (Collins, 1975).

In oilfields being produced by waterflood, the brine produced with the oil is aerated to complete the oxidation and precipitation of oxides and then reinjected into the oil-producing formation to displace the oil to production wells. Thus, the injected water is completely saturated with oxygen when it arrives in the petroleum reservoir.

Ivanov and Belyaev (1983) examined the microbial flora of a water injection well and found that bacterial oxidation of oil took place in the zone of contact between the injected low-salinity waters and stratal waters of the oilfield. The products of

Pennsylvanian	Mississippian	Devonian	Silurian	Ordovician	Cambrian
7	9	50	37	20	17
43,000	41,500	48,000	49,100	31,000	23,400
170	430	3,100	1,900	990	440
0.55	1	4	4	2	3.3
0.15	0.40	0.4	0.4	0.2	0.6
9,100	8,900	18,000	21,000	6,100	4,000
1,900	1,600	2,900	4,300	1,300	1,300
600	630	1,000	730	340	125
30	5	40	15	6	–
15	40	30	30	20	7
60	12	175	–	56	–
–	3	2	–	–	–
87,000	85,000	115,000	122,000	62,000	46,100
490	410	1,060	520	300	520
210	110	30	17	25	18
130	185	155	115	270	260
40	450	30	–	25	–
430	540	450	830	1,070	1,170
430	370	130	90	520	30
300	210	110	80	140	60

aerobic oil destruction stimulated the activity of methane-producing bacteria. They also examined the distribution of microorganisms and the rate of bacterial methanogenesis in the flooded oilfield and found that the population of microorganisms and the rate of methanogenesis increased as stratal waters were freshened by the injected low-salinity brines. This correlated with the decrease of organic carbon and the increase of bicarbonate concentration, as well as with methane isotopic ($\delta^{13}C$) composition becoming less and the bicarbonate carbon isotopic ($\delta^{13}C$) composition becoming greater. Thus, the process of bacterial oxidation of oil took place in the zone of contact between the injected fresh waters and the stratal waters of the oilfield.

Just as many oilfields, however, are being produced by other mechanisms that do not require the injection of brine into the oil-producing formation. In these fields, the produced brine is a waste product which is disposed of by injection into subsurface formations from which no fluids are withdrawn.

Oilfield brines have a wide range of concentrations of a large variety of cations and anions. General concentrations of oilfield produced brines range from less than 100 mg/l to more than 200,000 mg/l. Collins (1975) listed the analyses of oilfield waters occurring in rocks of various geologic ages (Tertiary, Cretaceous, Jurassic, etc.; Table 2-I). Concentrations of ions from waters of different geologic formations and, in places, even at different areas of the same formation differ widely. Due to the wide ranges of salt concentrations, there is no typical oilfield brine. Most of the produced brines have sodium or calcium chlorides as their major constituents, but magnesium, bicarbonate, and sulfate predominate in some. Other ions are present as minor constituents, and in some brines ions such as bromine and iodine are present in commercially economic quantities. According to Collins (1970), the highest concentrations of minor cations and anions that may interfere with microbial systems are (expressed in mg/l): lithium—400, barium—670, boron—450, bromine —6000, and iodine—1400.

Compared to the composition of modern sea water (Table 2-I) and assuming that the original brines were approximately the same as modern sea water, the subsurface waters have undergone considerable change. Some waters were diluted by percolating meteoric water or by compaction fluids, i.e., water expelled from muds and shales as they compacted during burial. The salt content of other waters was probably increased by solutions of ions from the rocks which were contacted by the water (Rieke and Chilingarian, 1974).

The present emphasis on microbial EOR is for injection of microbial cultures to improve the recovery of oil from oilfields undergoing active waterflooding or in relatively shallow deposits of heavy oils. In the first case, the oilfield brine can be adjusted by injection of low-salinity surface water and the water will be saturated with oxygen. Thus, the adaptation of the microbes to oilfied waters or the surface treatment of injection waters for microbial growth must be carefully considered before live cultures are injected. Some bacteria of genus *Clostridium* and *Bacillus*

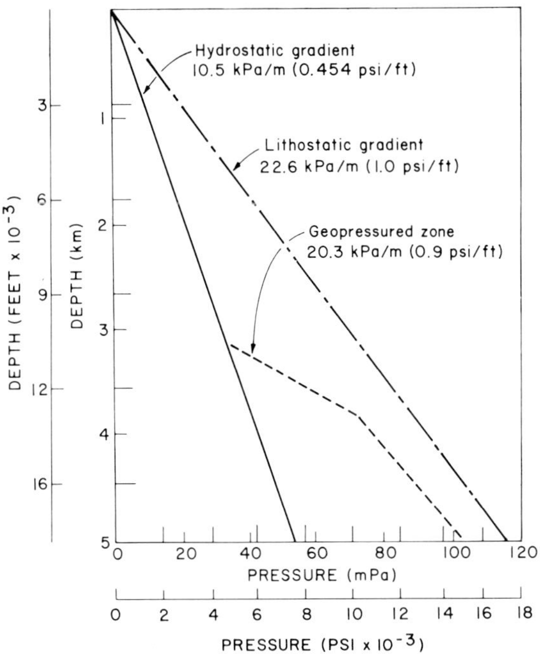

Fig. 2-1. Subsurface pressure gradients (Donaldson, 1980).

have been isolated that can tolerate 50,000 mg/l of sodium chloride; however, greater salt tolerance may be necessary for practical field applications (Grula, 1982; Knapp, 1982). The problem of salt tolerance is not so great for heavy-oil deposits, because most of these are less than 1000 m in depth and contain low-salinity brines, probably because of contact with meteoric water.

SUBSURFACE TEMPERATURES AND PRESSURES

The hydrostatic pressure gradient of subsurface formations is normally calculated by assuming that pressure is àpproximately equivalent to the hydrostatic head of a column of water. This approximation is useful when no specific data is available, but the concentration of salts, the presence of gas in subsurface formations, and zones of abnormally high pressure within confined aquifers all have a major influence on the actual pressure existing in a specific reservoir. Fig. 2-1 shows two pressure gradients within which most of the pressures associated with hydrocarbon deposits lie. The hydrostatic pressure gradient is 10.5 kPa/m (0.454 psi/ft), and at the other extreme is the lithostatic pressure gradient which is about 22.6 kPa/m (1.0 psi/ft).

Anomalies in the hydrostatic pressure gradient are caused by the presence of dissolved gases, sharp decreases or increase in the total dissolved solids, and

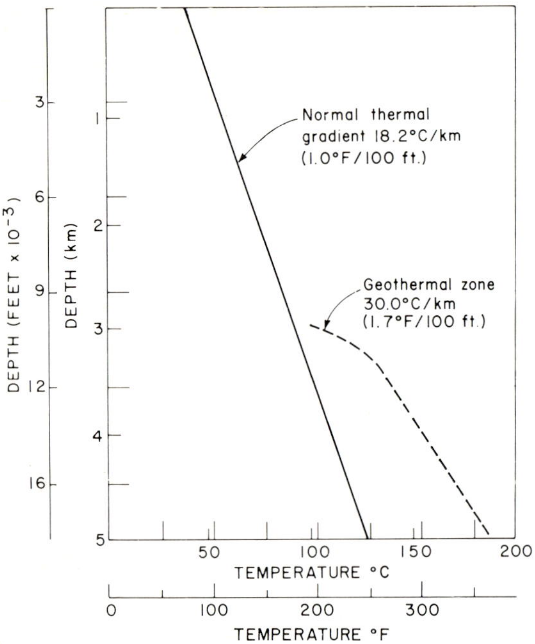

Fig. 2-2. Subsurface temperature gradients (Donaldson, 1980).

temperature variations in the subsurface. Although abnormally high pressures are encountered in some reservoirs, e.g. in the geopressured fairway of the Gulf Coast of the United States, at a depth greater than 3000 m, the pressure gradient in these zones only increases to about 20.3 kPa/m (0.9 psi/ft), because the lithostatic pressure cannot be exceeded without fracture of the formation and relief of the high pressure.

The normal hydrostatic pressure gradient can be expected in petroleum reservoirs that have long been under waterflood operations and are candidates for microbial EOR.

The normal geothermal gradient follows a similar trend to the pressure gradient increasing with depth, as illustrated in Fig. 2-2. Beginning at the mean surface temperature, it increases at approximately 18.2°C/km (1.0°F/100 ft) of depth. Thus if the mean surface temperature is 26.7°C (80°F) and the depth of the reservoir is 1524 m (5000 ft), one would anticipate a temperature of approximately 54.4°C (130°F), and referring again to Fig. 2-1, a pressure of 16.0 MPa (2270 psi).

If the oilfield has been waterflooded for a long time, there will be a temperature gradient within the formation extending radially from the cooler water at the wellbore into the formation until the normal reservoir temperature is attained.

The temperature gradient exists due to transfer of heat from the deep mantle. Abnormalities in heat transfer of formations or differences in the heat transfer coefficients of different formations due to changes in lithology, are responsible for the deviations from the normal geothermal gradients. Thus, some formations may have higher temperatures than others, because an overlying formation has a lower heat transfer coefficient and acts like a "blanket" (Tissot and Welte, 1978):

heat from the mantle $= 2 \times 10^{-6}$ cal/cm^2 s

thermal conductivity of sedimentary formations = 3–12 cal/cm^2 s °C

In considering the feasibility of using microbial colonies within a petroleum reservoir to enhance oil recovery, tests should first be conducted with the microbes at simulated reservoir conditions in a sedimentary rock to determine if the microbes will grow and produce the desired metabolic products at the conditions of the target reservoir. The microbes must remain biologically active at elevated temperature and pressure, and they must be able to tolerate the high electrolyte concentrations of oilfield brines. The petroleum reservoir is completely devoid of oxygen, except for a narrow zone around injection wells, and in many cases the oilfield brines contain metals, bromides, and iodides that may influence their metabolic activity when subjected to the elevated temperatures and pressures. Thus, an anaerobic or facultative bacterium, which is a suitable candidate for EOR at ambient conditions, may be inhibited in growth and/or metabolic activity, causing it to diminish its production of desirable metabolic products or to produce entirely different ones that are

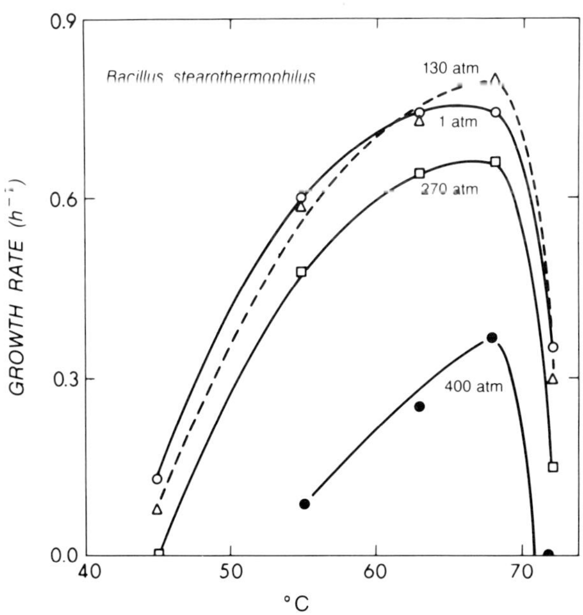

Fig. 2-3. Effects of pressure and temperature on the growth of *Bacillus stearothermophilus* in tryptone-glucose-Marmite broth plus 1% (w/v) KNO_3 (from Marquis, 1983, reproduced with permission).

ineffective for EOR. Bubela (1983) observed that the morphology of microbes changes when they are subjected to high pressure, changing from rod-shaped to coccoidal form. They burst when depressurized suddenly. If the pressure is released gradually, however, they assume rod shape again without apparent permanent effect. This change in morphology may precede a change in biological activity. Bubela (1983) also observed that *Desulfotomaculum denigrificans* increases its normal metabolic rate of growth by 10 times (as indicated by production of hydrogen sulfide) when grown at 70°C and a pressure of 10 MPa (1450 psi). Unfortunately this may not be true of other microorganisms with more desirable EOR biological activity, because the enzymes involved in cell metabolism may be adversely affected in the extreme environments presented in the subsurface.

This is an area where there is a severe lack of experimental observations; therefore, laboratory tests at the conditions anticipated for the target petroleum reservoir should be conducted. Donaldson et al. (1979) presented a detailed description of apparatus and procedures for conducting static and dynamic flow tests at reservoir conditions of brine and oil saturations, and elevated temperature and pressure.

Marquis (1983) studied the effects of elevated temperature and pressure on microorganisms in a limited number of tests made with sulfate reducers and *Bacillus stearothermophilus*. Low pressure, up to about 13.8 MPa (2000 psi), increases biological activity and even increases the optimal growth temperature by a few degrees (Fig. 2-3). An increase in pressure beyond some critical optimal point, however, results in inhibition of cell growth and metabolism at any temperature. Conversely, Marquis (1983) noted that microorganisms exhibit greater tolerance of high pressure at or near their optimal growth temperature.

PETROLEUM

Petroleum, which has been found in subsurface sedimentary geologic formations throughout the world, is a complex mixture of hydrocarbons frequently containing compounds of sulfur, nitrogen, oxygen, vanadium, and nickel. Petroleum is a very broad term that implies hydrocarbons found in the subsurface which may occur as gases, liquids, and even solid asphaltic deposits. The gases, which are frequently associated with oil reservoirs, are generally mixtures of the saturated hydrocarbons ranging from methane to butane. The crude oils contain thousands of compounds in a homogeneous mixture of such broad classifications as paraffins (straight- or branched-chain hydrocarbons), naphthenes (cyclic saturated compounds), olefins (straight or branched, containing one double carbon–carbon bond), aromatics (unsaturated cyclic compounds), polynuclear aromatics (fused unsaturated cyclic compounds), asphaltenes (complex polynuclear-chain structures), and compounds combined with other elements or metals. These broad groups of compounds vary greatly in quantity and distribution from one petroleum deposit to another. Petroleum which has been exposed to the atmosphere is deficient in low-molecular-weight compounds and paraffins due to evaporation and microbial degradation of the paraffins and other low-molecular-weight compounds. Microbial degradation is also evident in any crude oil that was exposed to meteoric water even though the depth of burial may be several thousand feet. The most prominent results of microbial attack are removal of low-molecular-weight hydrocarbons, especially the paraffins, and drastic increase in the petroleum viscosity and density. Prominent examples of these are heavy (high-density, high-viscosity) oil deposits in the San Juaquin valley of California, the Orinoco River belt in Venezuela, and the Athabasca tar sands of Canada (Chilingarian and Yen, 1978).

The fact that microbial degradation of petroleum occurs in the subsurface reservoirs and presence of various types of microorganisms (facultative and anaerobic) that have been isolated from petroleum reservoir (e.g., see Grula, 1982), is conclusive evidence that a petroleum reservoir is not a sterile environment. The naturally occurring microbial activities are detrimental to petroleum deposits from the standpoint of commercial value of the oil, because they result in loss of the most useful hydrocarbons. The fact of their occurrence, however, is sufficient incentive for specific research to develop microbial cultures that can produce desirable effects in the subsurface reservoir. A massive microbial culture that produces gases and solvents (such as the genus *Clostridium*) could bring about release of oil from the sand, decrease of oil viscosity by gas solution, and pressurization of the reservoir. The procedure to bring this about might entail a fermentation period of the reservoir during which all production is stopped. Microbes that produce polymers might be used in the subsurface to plug zones of high permeability and, thus, increase the effectiveness of oil displacement by injection of water. Alternately, the products of microbial metabolism developed in fermentation vessels on a large scale may be used for oil displacement (altering the viscosity and wettability of heavy oils), for pumping and transportation, and for cleaning tanks.

WATERFLOODING

The most probable targets of microbial enhancement of oil recovery (MEOR) are reservoirs that have reached the limits of oil production by injection of water to displace oil to producing wells. Fig. 2-4 is an idealized sketch of the situation that exists when a field reaches the economic limit of oil production from the injection of water (brine from the producing formation and other sources) into the center well of a "5-spot" configuration to displace the reservoir oil to the four corner wells. The pattern, or some similar arrangement of injectors and producers, is repeated throughout the field.

A single term encompasses the parameters that govern the pattern of residual oil remaining in a reservoir after attempts to produce the oil are abandoned. That term is "areal sweep efficiency". A 100% sweep efficiency can only be obtained if the

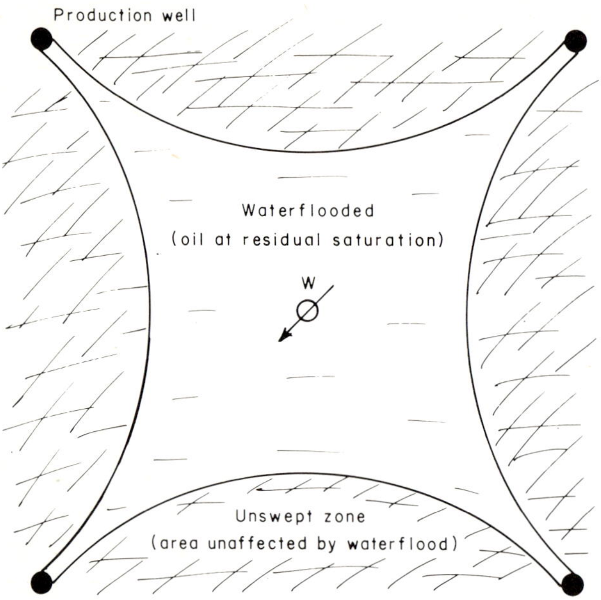

Fig. 2-4. Idealized sketch of a "5-spot" configuration with an injection well in the center and four corner producing wells. Areas of bypassed oil are in the unswept zone.

displacing fluids enter and exit from a homogeneous medium at openings exposed to the entire cross-section of the porous medium. This is attainable only in the laboratory for fluid displacement tests with cores. In the field, the displacing fluid enters through a relatively small hole and the produced fluids are withdrawn at some distance from a similar point source. Consequently, the area contacted by the displacing fluid as it moves from the injection to the producing well is considerably less than 100% at breakthrough (when the displacing fluid reaches the exit well). The problems faced by the oilfield operator are how to determine the areas bypassed by the displacing fluid, and how to produce these bypassed areas. The sweep efficiency of various well location patterns and fluid mobility ratios have been studied with numerical, fluid flow, and electrical models of various types, and many of them are summarized by Donaldson and Crocker (1977).

During the displacement of oil or gas, a front develops which is characterized by a high saturation of the displaced oil or gas, followed by a fairly sharp decrease in saturation behind the front. Mathematically, the mobility ratio, M, is defined as the effective mobility of the displacing fluid behind the front (k_{rw}/μ_w) divided by the effective mobility of the displaced fluid ahead of the front (k_{ro}/μ_o):

$$M = \frac{k_{rw}/\mu_w}{k_{ro}/\mu_o} \tag{2-1}$$

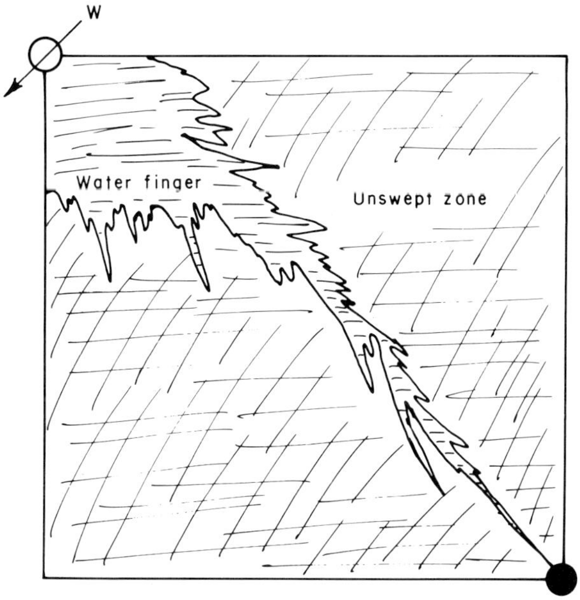

Fig. 2-5. One quadrant of a "5-spot" configuration showing a finger of water extending from the injection well to the producing well early in the life of the waterflood.

Mobility ratios greater than 1.0 are regarded as unfavorable resulting in frontal instability and channeling of the aqueous phase, whereas mobility ratios less than 1.0 have the opposite effect and are, therefore, considered as favorable.

Good, very economically favorable, displacement of oil occurs if the oil viscosity is low. The closer the oil viscosity is to the viscosity of the water, the more efficient is the displacement of oil, or "sweep efficiency" leaving a smaller unswept area of residual oil when waterflooding is no longer economical. On the other hand, if the viscosity of the oil is much greater than that of the water, the injected water "fingers" through the oil zone to the producing well early in the development of the waterflood, as shown in Fig. 2-5. As soon as water breaks through the oil zone to the producing well, this finger of water acts like a straight pipe, or channel, for transfer of water from the injector, and the water-to-oil production ratio increases rapidly until it is no longer feasible to attempt production of oil by waterflooding.

The growth of "fingers" of water and water channels from the injection well to the production well is influenced by the wetting characteristics of the oil–water–rock system. Wettability is a factor when the reservoir is oil wet, because this condition promotes fingering of the aqueous phase (Donaldson and Crocker, 1977).

The porous rock matrix of a petroleum reservoir is never completely homogeneous, as depicted in the idealized sketch of Fig. 2-4. In fact, there are numerous zones of high permeability that act as conduits for almost direct transfer of water from the injection well to the producing well resulting in the same condition that can develop due to an unfavorable viscosity ratio between the oil and water. Many methods have been used, and new ones continually are suggested, to plug these zones of high permeability (injection of polymers, gels, fine solids, etc.). The potential exists for discovery of bipolymers that are far more effective for plugging high-permeability zones than are the organic polymers and gels currently being used. In addition, it may be possible to sustain the growth of polymer-producing microbes within the high-permeability zone of a petroleum reservoir. In this case, the bacteria would tend to continually travel into the zones of higher water flow rates plugging these as soon as they develop and, thus, maintaining the flow of water within the low-permeability zones, which are commonly bypassed by the injected water. Such a culture developed within a more homogeneous reservoir with a favorable oil-to-water viscosity ratio would perform a similar function, restricting the flow of water in the "watered-out zones" as shown in Fig. 2-4, and increasing both the sweep efficiency and production of "residual oil".

Bacteria that produce polymers, which lower the interfacial tension between water and oil, may serve in a dual role: (1) plugging the high-permeability zones, and (2) enhancing oil production by surfactant actions that displace oil trapped in cul-de-sac pockets and areas isolated by narrow pore throats.

Anaerobic bacteria that produce gases and solvents will exert other favorable chemical and physical effects. For example, the production of carbon dioxide or methane will bring about reduction of the oil viscosity by solution of the gas in the oil and will increase the pressure in the oil phase, causing displacement from trapped capillaries and aiding in mobilization of the oil to the producing wells. The production of acid will improve permeability of the reservoir by dissolving parts of the rock matrix. The production of solvents, such as alcohols and ketones, will aid in displacement of the oil by dissolving and emulsifying actions.

Yarbrough and Coty (1983) reported on an experiment in Arkansas, U.S.A., where they used *Clostridium acetobutylicum* to obtain increased production from a well that had reached its limit of economical production. They reported that during the period of the test (approximately 1 year) the cumulative amount of carbon dioxide produced from the single well where it was measured was 5700 m^3 (11,000 kg) and that the measured amount of produced acid from the same well was 35,000 kg. They did not measure the production of alcohols and ketones, but the actual production of that much metabolites alone indicates that treatment of a reservoir with microbes holds the promise for considerable enhancement of oil recovery. Inasmuch as the production of carbon dioxide and acids was measured at only one well, one must conclude that the bacteria actually produced at least four times as much metabolites. Yarbrough and Coty (1983) reported a two-and-one half increase of production from the single well.

Another procedure that may be even more effective for production from a waterflooded area would be to close the production wells for several months while a fermentation process is carried out deep within the petroleum reservoir by injection of bacteria and nutrients. This would allow time for repressurization of the reservoir and solution of the gases in the oil and greater dispersion in the reservoir, rather than channeling the metabolites directly to the production wells.

Another approach to increasing production of watered-out wells was suggested by Johnson (1979) at the August 1979 DOE Forum in San Diego, California, U.S.A. His method involves the use of gas-solvent-forming bacteria to clean out the pores of the producing wells by injecting the bacteria and nutrients into the formation around the wellbore, allowing time for maximum fermentation to take place, and then backflowing, or pumping out, the production well. The production of gas and solvents and increase of pressure near the wellbore will effectively clean out these pores to allow better flow of oil into the producing well. According to Johnson (1979):

(1) A mixed bacteria culture of *Bacillus* and *Clostridium* is used.

(2) Time required for fermentation in the subsurface is 10–14 days.

(3) Cattle feed molasses is used as the fermentable nutrient.

(4) The copious amounts of CO_2, CH_4, and N_2 partially repressurize the reservoir.

(5) The low-molecular-weight acids react with reservoir carbonates to form additional CO_2.

(6) The solvents formed reduce the oil-water interfacial tension.

(7) MEOR is most effective for oil between 15° and 30° API.

(8) Carbonate reservoirs or those that have carbonate cementation respond better to microbial treatment.

(9) The rate of fermentation begins to decrease when the salt concentration of the formation exceeds 10%.

(10) The single-well "huff-and-puff" process will significantly reduce formation damage around the wellbore.

(11) Additional recovery of 20–35% of the oil-in-place may be expected.

The discussions of Yarbrough and Coty (1983) and Johnson (1979) are encouraging; however, at this time, there are no well-documented, detailed, field tests in the literature that discuss the entire MEOR process, i.e., the procedures used to develop a culture of microbes, a quantitative discussion of field processes, the reservoir geology, oil and brine characteristics, and the production history before and after the application of MEOR. Any field use of microbes, therefore, is still largely an experimental venture, scientifically, economically, and from the standpoint of engineering.

RESERVOIR ROCKS

Virtually all rocks that contain commercial deposits of hydrocarbons are porous sedimentary rocks. In a few cases highly fractured igneous and metamorphic rocks have been formed into traps for oil accumulation by tectonic events, but these are rare and, therefore, their characteristics are not discussed. Sedimentary rocks are composed of fragments of other rocks derived from mechanical and chemical deterioration of igneous, metamorphic and other sedimentary rocks. Mechanical weathering produces boulder size rocks, gravel, sand grains, silt and clay. These fragments may remain in the local area or they may be transported by winds and water to other sites to enter into the formation of conglomerates, unconsolidated sand deposits and sandstones. Chemical weathering may occur simultaneously with mechanical weathering which accelerates the process. Mechanical weathering provides access to large area for contact by water solutions containing acids such as carbonic acid, and hydrates of some minerals. The result is the formation of various clay minerals, silica and carbonates (Chilingar et al., 1967).

The composition of sand and sandstone sediments vary widely because numerous factors are involved in their formation: the sources of the detrital materials, mechanism of transportation, depositional environment, age and diagenesis. The four main sources of sand are (1) plutonic rocks, which supply mostly quartz and feldspar; (2) terrigenous sediments, which supply mostly quartz and rock fragments; (3) volcanic rocks, which supply mostly rock fragments and glass; and (4) carbonate rocks, which supply carbonate debris. The depositional environments in which sediments form are very important in determining (a) regional distribution; size, shape, and thickness of the sandstone bodies; (b) stratigraphic relationships; (c) grain-size distribution; (d) mineralogic composition; and (e) depositional structures. The following are the most important environmental sandstone types: (1) fluvial and alluvial, (2) beach, (3) bar, (4) dune, (5) barrier island, (6) erosion plain, (7) desert, (8) intermontane basin, (9) lake, (10) glacial, (11) volcanic terrain, (12) marine shelf, (13) tidal, (14) continental slope (e.g., turbidites), and (15) bathyal.

Pure quartz sands are common and quartz grains make up a sizable portion of many sands, especially in older, mature rocks that may have been reworked several times (multi-cycle). Quartz is a hard, durable mineral, that withstands weathering far better than most of the minerals with which it may have been associated in the original igneous source rock. The absence of heavy minerals in sands could be due to the severity of weathering and length of exposure to an oxidizing or leaching environment regardless of age and/or intrastratal solution. Abundance of feldspars in many older sands could indicate that the sediments were exposed to rapid erosion, a short or rapid transport and deposition, and little or no reworking. Inasmuch as the softer, less durable rocks are often reduced to smaller sizes and constitute smaller fractions of the total sand composition, their presence is reflected in the textural features of sands.

Arkose and greywacke sandstones (Fig. 2-6) are rarely composed only of mineral grains with associated intergranular pore space. The pore space usually contains some silts, clays and various chemical cements. The interparticle material may be no more than a fine-grained expression of the larger mineral grains, which may or may not have been mechanically and/or chemically altered into other forms such as clay. Orthoquartzites, on the other hand, usually contain only cement.

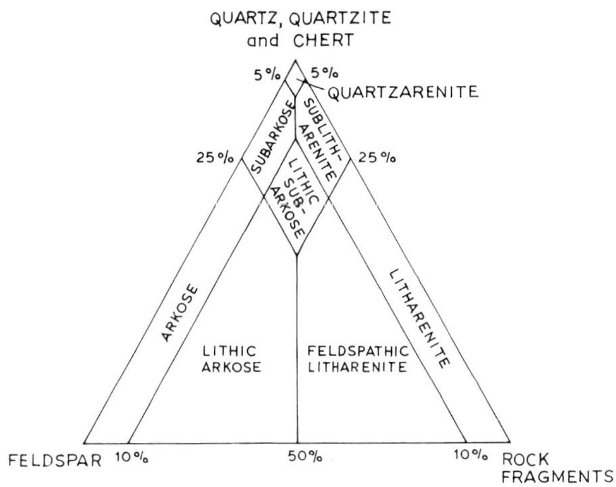

Fig. 2-6. Classification of sandstones illustrating the mixed composition of sedimentary rocks. The descriptive terms are used for general characterization of rocks (McBride, 1963, p. 667, fig. 1).

There are numerous sand classifications, most of which are basically methods of determining the relative proportions and composition of fragments and matrix. The descriptive adjectives referring to composition include: arenaceous, argillaceous, calcareous, feldspathic, ferruginous, quartzose, and quartzitic. These terms are then combined with a textural term into an expression such as "arenaceous siltstone" (Chilingarian and Wolf, 1975).

Sandstones are the cemented grains of silica and other rock fragments which are held together with deposits of carbonates, silica, clay, and minor quantities of many other minerals. The interconnected pore space may range from less than 1% for highly compacted deep sandstones to 30% for surface outcrops that are not subjected to an overburden pressure and are exposed to flow of meteoric water. Frequently, a secondary period of cementation had occurred later in the period of existence of the sandstone where solutions containing gypsum, calcite, glauconite, siderite, etc., entered the sandstone, and these cements were deposited in the pores resulting in decreased permeability, porosity, and increased hardness of the sandstone.

The carbonate rocks generally originate from chemical precipitation or biologic concentration followed by precipitation. These deposits, although quite massive, usually contain inclusions of detrital materials. The porosity may be quite variable because some of the porosity was derived from solution after deposition of the rock. Tectonic movements produce fractures that allow percolation of water which enlarges the fractures, produces solution vugs, and precipitates minerals that may act as cements for loose grains, and filling (healing) of fractures; thus the processes result in a complex pore geometry. Reaction of the calcium carbonate rock with magnesium in the percolating waters results in conversion of calcite to dolomite, $CaMg(CO_3)_2$, and creates porosity in many carbonate rocks (Chilingar et al., 1979).

Seven different forms of calcite—$CaCO_3$, magnesite—$MgCO_3$, and dolomite—$CaMg(CO_3)_2$ have been recognized: (1) calcite; (2) magnesite; (3) dolomite; (4) arogonite (5) solid solution between calcite and dolomite, such as skeletons of organisms, (6) nesquehonite; and (7) hydromagnesite. The latter two forms are rare, but have been shown by Frolova (1955) to be present in the carbonate–sulfate rocks of the Kuybyshev area of the USSR. Details of carbonate mineralogy are provided by Graf (1960).

Shales are the result of compacted muds and are of little importance as petroleum-containing materials, because they are very fine-grained and have low permeability (≈ 1 mD) and low porosity. Their importance is in their function as source rocks for the genesis of petroleum and as impermeable barriers above and below petroleum reservoirs. Shale lenses also are inclusions within reservoirs that impede the symmetrical transport of injected fluids from injection to production wells. Shales exhibit plastic flow behavior during tectonic loading and motion. Shales frequently dilute the subsurface brines by discharging stored low salinity water into the brine aquifers when subjected to tectonic stress.

Two of the most important properties of fluid-filled subsurface reservoirs are their porosity and permeability. The porosity is defined as the ratio of the interconnected pore volume to the bulk volume of the rock. Small laboratory samples are frequently inadequate for a total description of the reservoir porosity, because the reservoir itself may have added pore space between blocks in fractured zones, large solution vugs (which are common in carbonate rocks), and extreme ranges of porosity, both vertically and horizontally. Engineering procedures and calculations, however, are based on (1) carefully measured laboratory samples, (2) the vertical profile of porosity obtained from well logs, and (3) the results of well test analysis. Together, these three methods yield sufficient data, to determine oil, gas, and water saturations, which are reliable for engineering evaluation of petroleum reservoirs.

The second important property of oil-bearing rocks is the permeability, which is a measure of the ability of a porous medium to transmit fluids. It is best expressed by its mathematical formula:

$$k = Q\mu L/Ap \tag{2-2}$$

where k = permeability in darcies, Q = volumetric rate of flow in ml/s, μ = viscosity in cP, A = cross-sectional area of flow in cm^2, p = pressure differential in atm, and L = length of flow in cm.

The darcy (or millidarcy) is the most common unit of permeability in the literature, but there is an increasing trend to express it as square centimeters (1 D $= 9.87 \times 10^{-7}$ cm^2).

Good examples of the porosity and permeability of unconsolidated sands, determined for the Wilmington oilfield, California, U.S.A., by Donaldson and Staub

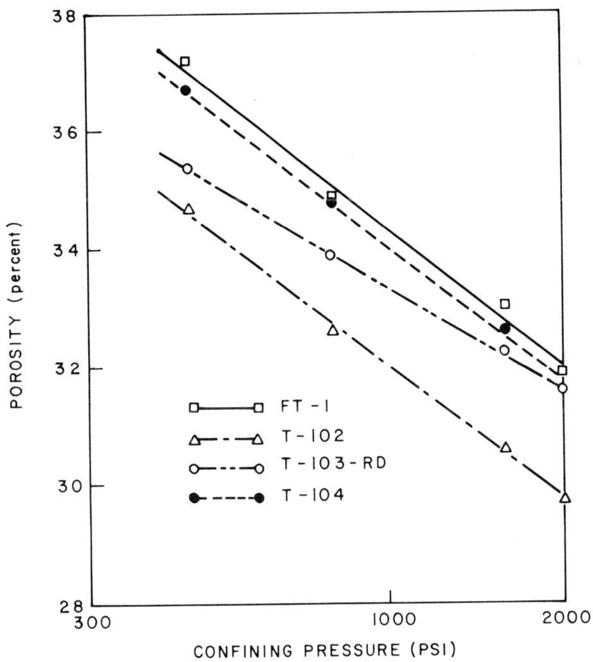

Fig. 2-7. Effect of overburden pressure on porosity; cores compacted dry (Donaldson and Staub, 1981).

(1981), are shown in Figs. 2-7 and 2-8. Due to the variability of reservoirs, generalizations of the magnitude of porosity and permeability are of questionable value; however, the porosity and permeability ranges shown in Figs. 2-7 and 2-8, and their decrease with increasing depth of burial (overburden pressure) are well within the ranges of many other formations composed of unconsolidated sands. Hence, if one is dealing with an unconsolidated sand formation, one might anticipate a porosity of 28–36% and permeability range of 400–1200 mD as a first guess.

Consolidated sands, or sandstones, are far more irregular than unconsolidated sands. A study of the porosity, permeability, and surface area of outcropping sandstones was conducted by Donaldson et al. (1975), and is summarized in Table 2-II. Several properties of sandstones are evident from the table. The Berea and Cottage Grove sandstones are very close in porosity, permeability, and density, yet the Cottage Grove Sandstone has a surface area and ion exchange capacity several times larger. Later studies have shown that although the quantity and types of clay are the same, the clay minerals in the Cottage Grove Sandstone are distributed along the pore walls, whereas in the Berea Sandstone they act more as cementing agents between the grains, and the pore walls are free of clay, thus exposing less clay for increasing surface area and ion exchange. The Coffeyville Sandstone has secondary cementation, which accounts for its much lower permeability and smaller pore sizes (Fig. 2-9). The Noxie Sandstone has larger grain sizes; hence, its pores are

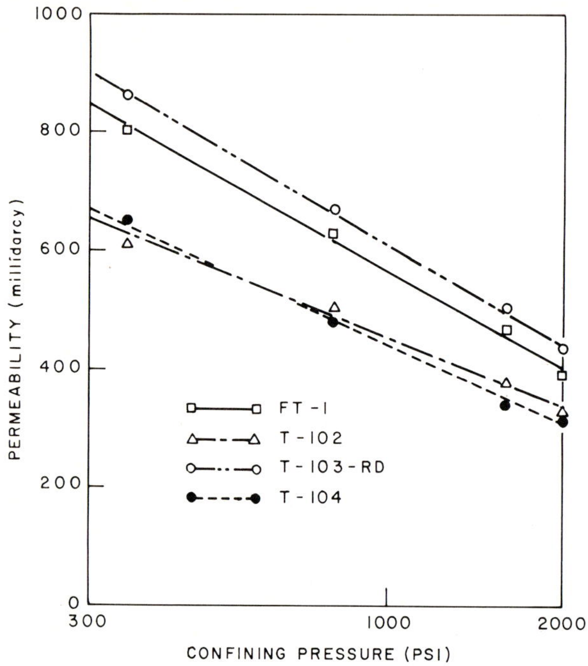

Fig. 2-8. Effect of overburden pressure on permeability (Donaldson and Staub, 1981).

larger in diameter (Fig. 2-9) giving rise to higher permeability and lower density. The Torpedo Sandstone has a large amount of distributed illite clay in its pores which accounts for the much larger ion exchange capacity.

Pore size distribution studies of the outcrop rocks (Fig. 2-9) also were conducted by Donaldson et al. (1975). The difference between the Berea and Cottage Grove sandstones is once more evident by the fact that the Berea Sandstone, with its clean pores, is skewed toward the larger pore sizes and the curve of the Cottage Grove Sandstone, with clay lining its pore walls, is skewed toward the smaller pore sizes. The Western Tight Gas sand from a depth of 3810 m (12,500 ft) has a bimodal pore

TABLE 2-II

Physical properties of outcrop sandstones obtained near Bartlesville, Okla., U.S.A., compared to the Berea Sandstone quarry outcrop from Amherst, Ohio, U.S.A. (Donaldson et al., 1975)

	Porosity (fraction)	Permeability (mD)	Density (g/ml)	Surface area (m^2/g)	Ion exchange capacity (meq/kg)
Berea ss	0.190	325	2.10	1.00	5.3
Cottage Grove ss	0.250	350	1.92	2.50	18.0
Coffeeville ss	0.220	60	2.10	3.00	–
Noxie ss	0.270	100	1.95	2.85	29.3

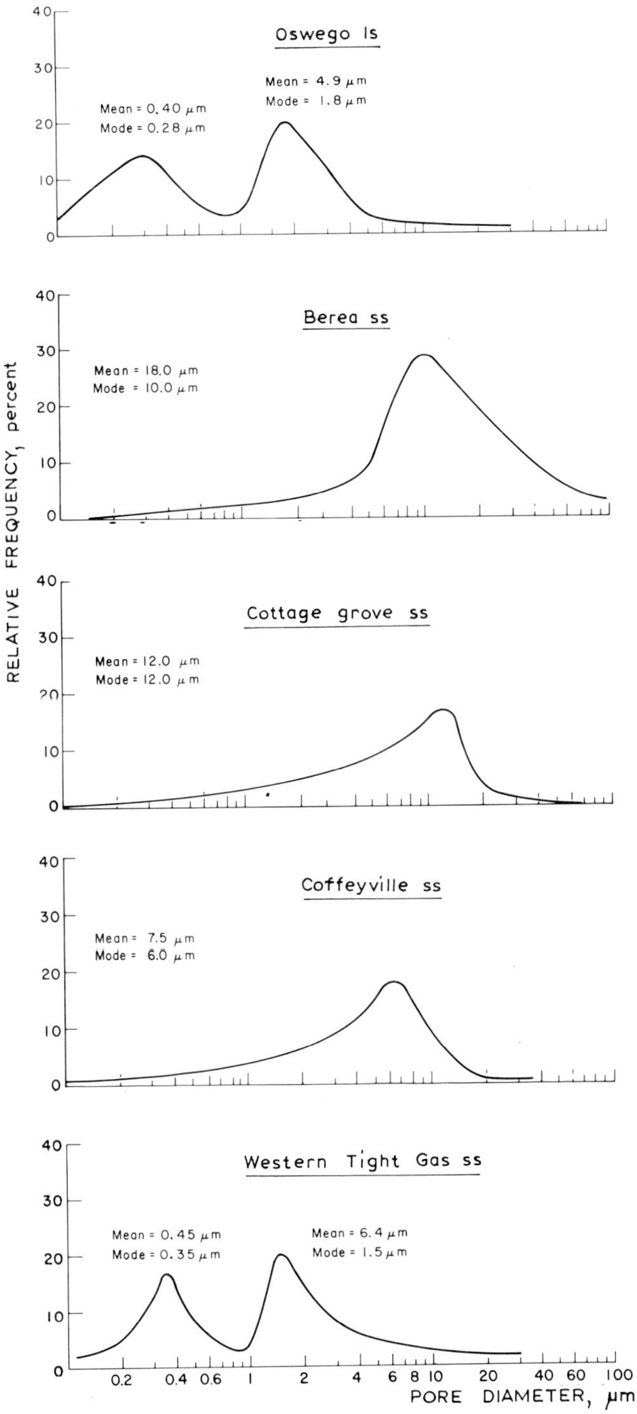

Fig. 2-9. Pore size distributions of several sedimentary rocks (Donaldson et al., 1975).

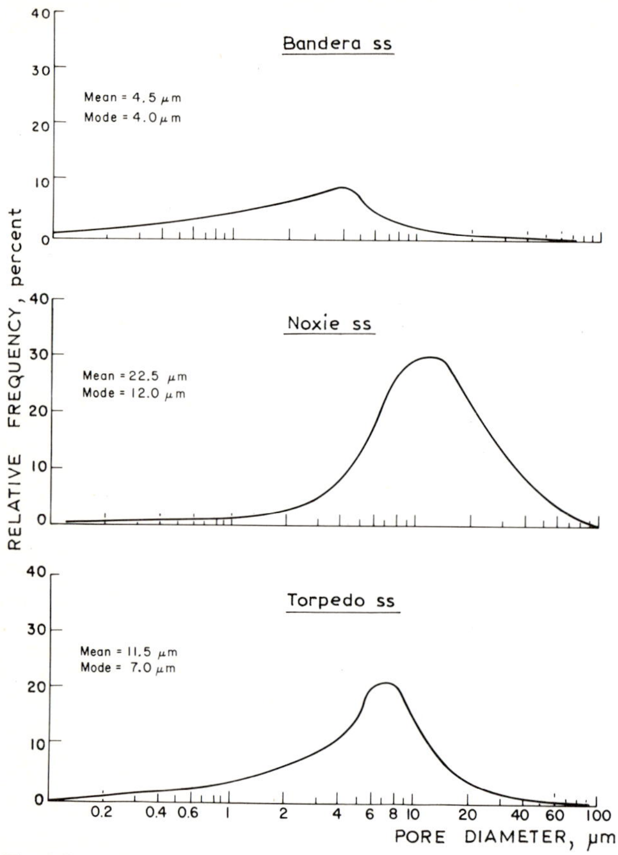

Fig. 2-9 (continued). See p. 33 for legend.

size distribution, which shows the effects of high overburden pressure on compaction of the sandstone. The grains have been squeezed together, with particles of clay interspersed, to such an extent that the pores existing in the surface have been reduced in size to 1.5 μm. Other studies showed conclusively that the smaller of the distributions (0.2 μm mode) was due to the development of microfractures after extraction of the core as a result of expansion of the rock upon relief of the overburden pressure.

Carbonate rocks are even more variable in composition and complexity of pore geometry than the sandstones. The biomodal curve and small pores of the Oswego Limestone, shown in Fig. 2-9, are not unusual, but within a few meters the pore sizes may be a thousand times larger because of the presence of solution vugs.

The Oswego Limestone (a brine aquifer south of Bartlesville, Oklahoma, U.S.A.) contains a high volume of brine at pressures sufficient to flow as artesian wells, but the small sample taken between solution vugs which is shown in Fig. 2-9 is almost a solid block with a bimodal distribution of very few small pores, which have not been widened by fluid flow of the brines that produced the solution vugs in the aquifer. Examination of the actual carbonate formation by extraction of cores from various wells, and well logs, is the only way to understand a specific carbonate reservoir. Even these tests should be corroborated by well test analysis, where carbonate reservoirs are concerned, because of the extremes of variation in porosity and permeability that may be present, as illustrated in the one example of the Oswego Limestone.

REFERENCES

Bubela, B., 1983. Combined effect of temperature and other environmental stresses on microbiologically enhanced oil recovery. In: E.C. Donaldson and J.B. Clark (Editors), *Proceedings, 1982 International Conference on Microbial Enhancement of Oil Recovery*. NTIS, Springfield, Va., pp. 118–123.

Chilingar, G.V., Bissell, H.J. and Fairbridge, R.W. (Editors), 1967. *Carbonate Rocks*. Elsevier, Amsterdam, 471 pp.

Chilingar, G.V., Zenger, D.H., Bissell, H.J. and Wolf, K.H., 1979. Dolomites and dolomitization. In: G. Larsen and G.V. Chilingar (Editors), *Diagenesis in Sediments and Sedimentary Rocks, 1*. Elsevier, Amsterdam, pp. 423–536.

Chilingarian, G.V. and Wolf, K.H. (Editors), 1975. *Compaction of Coarse-Grained Sediments*. Elsevier, Amsterdam, Chapter 1.

Chilingarian, G.V. and Yen, T.F., 1978. *Bitumens, Asphalts and Tar Sands*. Elsevier, Amsterdam, 331 pp.

Collins, A.G., 1970. Finding profits in oil-well waters. *Chem. Eng.*, 77/20 (Sept. 21): 165–168.

Collins, A.G., 1975. *Geochemistry of Oilfield Waters*. Elsevier, Amsterdam, 496 pp.

Donaldson, E.C., 1972. Injection wells and operations today. *Am. Assoc. Pet. Geol., Mem.*, 18: 24–46.

Donaldson, E.C., 1980. Underground disposal of brines from geopressured reservoirs. *Proc., 73rd Annu. Meet., Am. Inst. Chem. Eng., Chicago, Ill., Nov. 16–20*, 30 pp.

Donaldson, E.C., 1982. The potential of petroleum microbiology. In: *Proceedings, 17th Intersociety Energy Conversion Engineering Conference, Los Angeles, Calif., Aug. 3–12*, 2: 1009–1014.

Donaldson, E.C. and Crocker, M.E., 1977. *Review of Petroleum Oil Saturation and Its Determination*. NTIS, Springfield, Va., BETC/RI-77/15, 43 pp.

Donaldson, E.C. and Staub, H.L., 1981. Comparison of methods for measurements of oil saturation. *Soc. Pet. Eng. Conf., San Antonio, Texas, Oct. 5–7*, SPE 10298, 12 pp.

Donaldson, E.C., Kendall, R.F., Baker, B.A. and Manning, F.S., 1975. Surface-area measurement of geological materials. *Soc. Pet. Eng. J.*, 15(2): 111–116.

Donaldson, E.C., Kendall, R.F., Pavelka, E.A. and Crocker, M.E., 1979. *Equipment and Procedures for Fluid Flow and Wettability Tests of Geological Materials*. NTIS, Springfield, Va., DOE/BETC/IC-79/5, 40 pp.

Frolova, E.K., 1955. Magnesite in Lower Permian deposits of Kuybyshev and Saratov Transvolga Region. *Izv. Akad. Nauk SSSR, Ser. Geol.*, 1955(5): 89–96.

Graf, D.L., 1960. Geochemistry of carbonate sediments and sedimentary carbonate rocks, Parts 1 and 2. *Ill. State Geol. Serv. Circ.*, 297: 39 pp. (Part 1); 298: 43 pp. (Part 2).

Grula, M.M., 1982. Microbial enhancement of oil recovery. *U.S. Dep. Energy/BETC Q. Progr. Rev.*, 29–32.

Ivanov, M.V. and Belyaev, S.S., 1983. Microbial activity in waterflooded oilfields and its possible regulation. In: E.C. Donaldson and J.B. Clark (Editors), *Proceedings, 1982 International Conference on Microbial Enhancement of Oil Recovery*. NTIS, Springfield, Va., pp. 48–57.

Johnson, A.C., 1979. In: *Conference on Microbiological Processes Useful in Enhanced Oil Recovery, San Diego, Calif., Aug. 29-Sept. 1*. NTIS, Springfield, Va., pp. 30–32.

Knapp, R.M., 1982. Microbial Enhancement of oil recovery. *U.S. Dep. Energy/BETC Q. Progr. Rev.*, pp. 29–32.

Larsen, G. and Chilingar, G.V. (Editors), 1979. *Diagenesis in Sediments and Sedimentary Rocks, 1.* Elsevier, Amsterdam, 579 pp.

Levorsen, A.I., 1966. *Geology of Petroleum*. W.H. Freeman, San Francisco, Calif., 2nd ed., 725 pp.

Marquis, R.E., 1983. Barobiology of deep oil formations. In: E.C. Donaldson and J.B. Clark (Editors), *Proceedings, 1982 International Conference on Microbial Enhancement of Oil Recovery*. NTIS, Springfield, Va., pp. 124–128.

McBride, E.F., 1963. A classification of common sandstones. *J. Sediment. Petrol.*, 33: 664–669.

Rieke, H.H., III and Chilingarian, G.V., 1974. *Compaction of Argillaceous Sediments*. Elsevier, Amsterdam, 424 pp.

Tissot, B.P. and Welte, D.H., 1978. *Petroleum Formation and Occurrence*. Springer-Verlag, Berlin, 263 pp.

Yarbrough, H.F. and Coty, V.F., 1983. Microbially enhanced oil recovery from the Upper Cretaceous Nacatoch Formation, Union County, Arkansas. In: E.C. Donaldson and J.B. Clark (Editors), *Proceedings, 1982 International Conference on Microbial Enhancement of Oil Recovery*. NTIS, Springfield, Va., pp. 149–153.

Chapter 3

THE POTENTIAL FOR IN-SITU MICROBIAL APPLICATIONS

GARY E. JENNEMAN

INTRODUCTION

It appears certain that no single method of tertiary recovery will be applicable for all reservoirs, and it is necessary to choose the best method for any given set of conditions. Thus, an arsenal of tertiary methods has been and is being developed, and all will be needed as our finite quantity of petroleum becomes more limited. Much work, however, remains to be done to design a predictable process for any given method under a given set of reservoir conditions.

A potential addition to this arsenal of methods will be the use of microorganisms and their metabolic activities as a tertiary recovery process. Only the in-situ use of microorganisms is considered here. Certainly, the ability of microorganisms to produce specific products in a controlled fermentation cannot be ignored in the search for additional tertiary methods of recovery. A good example of this power is the production of xanthan gum, which is the result of aerobic metabolism, carried out under carefully controlled conditions for the purpose of producing the polymer, xanthan. The product is then transported to the well and injected for a "conventional" polymer flood. Much of the injected polymer is adsorbed on the formation before it reaches the position in the reservoir where it will do the most good. Thus, more of the expensive polymer is utilized than would actually be needed if the material could be more accurately placed in the reservoir to achieve maximum results. To this must be added the transportation costs to deliver the material to the wellsite. Even if the material were manufactured at the wellsite (which could also be very costly), the need for excess material increases the cost of the method. The injection of microorganisms into the reservoir, along with or followed by an inexpensive food material which could be obtained locally, would lead to a more efficient, less costly process for oil recovery. Less material for the release of oil would be needed because it would be manufactured within the reservoir where the maximum effect would be achieved.

In the in-situ microbial enhanced oil recovery (MEOR), the reservoir becomes the fermentor for the production of the oil-releasing material. In an above-ground fermentation, a microorganism is chosen for its ability to produce the material sought. The microorganism may be manipulated to increase its productivity, and the conditions of growth are manipulated and carefully controlled for maximum production. Once these conditions have been met, a carefully controlled process is carried out to manufacture the desired product. In considering the in situ production of a microbial product, the conditions of growth are dictated by the physical and chemical characteristics of the reservoir to be treated. Little can be done to alter these conditions, so the major problem is finding the proper microorganism to economically produce the desired product under the conditions already existing in the reservoir. It is also necessary to examine which type of microbial activity would achieve the maximum effect in enhanced recovery. Although these may seem to be formidable problems, the knowledge and technology of microbiology as well as petroleum engineering has progressed to the point that solutions to these problems are attainable. As the need for increased tertiary recovery activity increases, it is likely that microbial processes will be added to the arsenal of tools available for application to selected oil reservoirs.

The purpose of this chapter is to examine the limitations of the application of microbial in situ processes based on the present state of knowledge.

HISTORICAL REVIEW

The concept of using microorganisms and their activities in an enhanced oil recovery procedure is not new. In 1926, Beckmann reported on the action of bacteria on mineral oil. As a result of his work, he proposed that bacterial enzymes could be used in oil recovery. His statement is as follows: "Without doubt the world's supply of oil is limited. It is, however, recognized that a large percentage of oil in the earth still remains after an oil well has stopped flowing, probably because the friction of the viscous oil against the sand is too great to permit the flow to continue. Now is it not possible that if a feeding medium inoculated with enzyme-producing bacteria were left in contact with the remaining oil, the enzymes produced might change the viscosity and gravity of the oil and cause a fresh flow?"

Although considerable information has been learned about residual oil and about microorganisms since 1926, his basic idea is still valid. He also suggested the application of microbial activities for treatment of bitumens and coal to break oil–water emulsions.

The first serious work on the utilization of microorganisms in oil recovery was done by ZoBell in the early 1940's in connection with American Petroleum Institute Project 43A on the role of bacteria in the origin of petroleum. In 1946, a patent was issued to ZoBell which involved the treatment of a petroleum-bearing formation with hydrocarbon-utilizing, sulfate-reducing bacteria to cause chemical and physical changes in the reservoir which would increase the production of oil. A second patent issued to ZoBell in 1953 extended the coverage of microorganisms used for oil release to include hydrogen-utilizing, sulfate-reducing bacteria. ZoBell (1947a, b, c) studied the mechanisms of oil release from sands by the action of bacteria with the following conclusion: "All that can be said regarding the practical use of bacteria to increase the recovery of oil from oil-bearing formations is that it appears to be a promising field for further investigation. Much more experimental work will have to be done by trained microbiologists in cooperation with petroleum engineers before any definite recommendations can be made."

ZoBell's pioneering work had considerable influence on subsequent developments on the use of microorganisms in oil recovery. A patent was issued to Updegraff and Wren in 1953 in which molasses is added to the reservoir as a substrate for growth of the sulfate-reducing bacteria. Other workers actively pursued the use of sulfate-reducing bacteria in enhanced oil recovery. The reader should consult Chapter 6 of the book written by Davis (1956) for details of this work.

Sulfate-reducing bacteria are detrimental to the production of oil from a petroleum-bearing formation. They produce H_2S which sours the oil or gas produced from the formation. They also contribute to increased corrosion problems and the subsequent formation of FeS as a result of their metabolic activities can cause non-selective plugging of a formation. Thus, the intentional injection and growth of these bacteria in a formation, even for the purpose of increased production, is subject to serious question. As a result, later work concentrated on the use of other types of bacteria for enhanced recovery processes.

Examples of work involving bacteria other than the sulfate-reducing bacteria are found in three patents issued to Hitzman (1962, 1965, 1967). He proposed using spores of bacteria and cells of aerobic and anaerobic species of bacteria. In these cases it was proposed that the petroleum itself would serve as food material for cellular activity. Since reservoirs are normally anaerobic, it would be necessary to inject oxygen or air to support aerobic growth. This has been proposed in patents issued to Hitzman (1956) and to Jones (1967).

The injection of any oxygen into a reservoir is subject to question from a practical viewpoint. The presence of oxygen contributes to corrosion problems, and in deep, hot reservoirs could lead to undesirable combustion of the oil. The injection of oxygen is currently employed in the in-situ combustion of reservoirs for the purpose of mobilizing heavy crudes; however, this adds to the cost of the process. Oxygen may be present in sufficient amounts in areas surrounding the wellbore region due to surface injection of water, but the transfer of oxygen into deeper regions of the reservoir would occur slowly, especially in high salt-containing brines. It appears to be more practical, therefore, to utilize anaerobic organisms for in-situ MEOR processes.

It is now generally believed that microorganisms cannot use petroleum components as food material under anaerobic conditions, at least within a time frame required for economical recovery. Moses et al. (1983) reported that anaerobic growth with petroleum as the sole carbon source can occur, but the rate of growth was slow requiring several months for detection. The various arguments on whether bacteria can grow under strict anaerobic conditions with petroleum components as the sole food source are not discussed here. This controversy can only be resolved with time and work, but it is crucial to the ultimate application of microorganisms in enhanced oil recovery processes. The general trend of thinking at this time is that it will not be possible, in most cases, to obtain adequate microbial activity in the absence of an added food material such as molasses. Any utilizable food material would be satisfactory provided the cost is not prohibitive.

Numerous patents and papers on various aspects of microbial enhancement of oil recovery have appeared since ZoBell's work in the 1940's. However, there was a noticeable drop in the number of reports in the mid-1960's. At that time, oil was in apparent abundance, and the price of oil was too low to justify the use of any tertiary recovery processes. Most of the petroleum microbiologists worked on other projects, including single-cell-protein production. The OPEC price increases and energy shortages in the 1970's reaffirmed the inevitable crisis that is facing mankind because of finite amount of oil remaining in the ground. There is more oil left in the ground in known fields than can be expected to be found in new discoveries. With the increasing price of oil, tertiary recovery processes seem more viable and the activities to develop such processes have increased. This led to a renewed interest in the potential for the use of microorganisms in the in-situ recovery processes.

One significant question was raised with the renewed interest in microbial enhancement. This pertained to the practical application of MEOR when considering the limitations of reservoir conditions on the growth of microorganisms. It will hopefully become apparent after reading the following sections that reservoir conditions vary a great deal from one to another. Since temperature, salinity, pressure, permeability, etc., can all pose limitations on microbial metabolism and growth, it would seem wise that when considering a reservoir for MEOR its potential must be evaluated based solely on its own physical and biological properties. In other words, a process that works fine for one reservoir may fail miserably for another unless all potential limiting factors are understood. There are, however, numerous reservoirs (large and small) that pose only minor limitations to at least some MEOR processes.

FIELD TESTS

Although the use of microorganisms in enhanced oil recovery is considered by most scientists in the area of enhanced recovery to be in an early developmental stage, numerous field tests have been run. Hitzman (1983) has reviewed the results from all field tests that have been reported in the literature. As shown in Table 3-I, over 200 field tests have been reported. The majority have been done in the United States, but these have practically all been treatment of stripper wells by a "huff and puff" method. The European tests have been more comprehensive, although it is difficult to obtain accurate information due to translation problems and due to

TABLE 3-I

Number of wells in various countries treated with microbial cultures (from Hitzman, 1983)

Cultures	Number of wells treated with cultures
Poland	17
Hungary	10
Czechoslovakia	6
U.S.S.R.	2
Romania	9
The Netherlands	2
U.S.A.	178

TABLE 3-II

Extremes reported in reservoir properties (from Hitzman, 1983)

Temperatures (°C)	22–97
Porosity (%)	11–36
Permeability (mD)	10–8100 (highly fractured)
Oil type	heavy asphaltic–light paraffinic
Rock type	sandstone–limestone

omissions in the original manuscripts. The dates of treatment range from 1954 to the present. Table 3-II gives the extremes reported in reservoir parameters and Table 3-III lists the cultures of microorganisms used as inoculum.

After a detailed examination of all the material reported in the literature, Hitzman (1983) formulated the following generalizations:

"(1) Microbial activity occurs in some reservoirs under some conditions and a positive oil response occurs.

TABLE 3-III

Cultures used as inoculum in different countries

Poland	mixed aerobic and anaerobic bacteria
Hungary	thermophilic mixed cultures
	sewage sludge
Czechoslovakia	sulfate-reducing and hydrocarbon-utilizing organisms
	sulfate-reducing bacteria
U.S.S.R.	mixed aerobic and anaerobic bacteria
The Netherlands	*Betacoccus dextranicus*
	slime-forming bacteria
Romania	adapted mixed enrichment cultures
U.S.A.	*Clostridium acetobutylicum*
	mixed cultures

(2) In-situ microbial growth can be stimulated by the addition of nutrients.

(3) Such microbial growth influences the properties of the fluids in the reservoir and the changes are both physicochemical and petrophysical.

(4) The flow patterns in the reservoir can be traced by following microbial growth patterns.

(5) Most of the mechanisms proposed for the MEOR process (such as CO_2, acid surfactant, etc., and changes in oil viscosity, gas production, etc.) have been shown to play a role in oil mobilization.

(6) Some reservoirs have conditions such as low permeability, salinity, pH, etc., that could limit microbial growth.

(7) The temperature and pressure conditions of some reservoirs may not be as restrictive as previously considered.

(8) Fracturing with microorganisms seems to be a successful technique in some cases.

(9) Even a small inoculum, or the stimulation of the native reservoir population, appears to exert an effect out of proportion to what would be expected from size alone.

(10) Anaerobic organisms seem to be preferred and more effective than aerobic forms.

(11) The technique employed to introduce the inoculum into the reservoir and their rate of penetration and action is dependent on the formulation characteristics and must be carefully considered to have a successful project.

(12) Better records and monitoring must be made to truly evaluate the success or failure of a test."

The results from the reported tests varied, and the relative success of each test is in the eyes of the beholder. In most cases, the production of oil was stimulated, but the degree of stimulation was not exciting by the standards now used to judge the success of a tertiary recovery process. It must be considered, however, that these tests were, for the most part, experimental and they were not performed under ideal conditions which would allow the proper choice of reservoir conditions. In most cases, the wells were stripper wells or from reservoirs that had been depleted to such an extent that additional treatment by traditional EOR processes known at the time of the microbial test were judged to be of little value. The open literature shows no case of a carefully planned, controlled field test being run in reservoirs with adequate properties to give a good evaluation of the efficiency of the process.

Perhaps the best documented field test is the 1954 Mobil test done in the Upper Cretaceous Nacatoch Formation in Union Country, Arkansas (Yarbrough and Coty, 1983). The microorganism used was *Clostridium acetobutylicum* which was fed in situ with a 2% beet molasses solution. The rate of oil production after treatment increased 250%. Unfortunately, this section of the reservoir had been watered out and residual oil saturation was 4.5–8.5%, so the actual amount of additional oil recovered was low.

RESERVOIR LIMITATIONS

A petroleum reservoir is composed of solid, liquid (aqueous and oleic), and gas phases, all of which can have enhancing, neutral, inhibitory, or lethal effects on the growth and survival of microorganisms. Toxic and inhibitory conditions found in these oil reservoirs could be detrimental to any microbial enhanced oil recovery process. It is the purpose of the following section to describe the physical and chemical nature of oil reservoirs and point out possible limitations to MEOR posed by oil reservoirs. Data are presented on physical and chemical characteristics for nine of the top oil-producing states in the United States as compiled by a computer search of both the Oily and ERDA data base files at the Office of Information Systems Programs (University of Oklahoma, Norman, Okla). Also presented is the potential for MEOR for thirty-five of the United States' giant oilfields.

Lithology

The solid phase of the reservoir is composed of various rocks and minerals. Oil is predominantly found trapped in sedimentary rocks but can also be found in other types of rock such as metamorphic and igneous (Amyx et al., 1960). Sandstones and carbonates (limestones and dolomites) compose the major classes of sedimentary rocks in which oil is found. In the data bases of those states searched, sandstone is the major class of reservoir rock reported, except in Kansas and New Mexico, where it was reported that more than 50% of the reservoirs contain predominantly limestone or dolomite (Table 3-IV).

Although silicates and carbonates pose little restriction on microbial activity, the adsorptive capacity of clays and some other minerals contained within the porous rock can interfere with biological processes. Under the proper conditions of pH and ionic strength, clays and rock possess charges on their surface that act to adsorb bacteria and inhibit their migration through the porous medium. Montmorillonite clays exhibit the greatest ion exchange capacity, whereas kaolinites are least adsorptive. Illites have intermediate exchange capacity. Clays can also adsorp water resulting in their swelling, thereby restricting microbial transport through the rock matrix. This swelling is reduced by the presence of salts (NaCl, KCl, $CaCl_2$) found in most reservoir brines. Veley (1969) as well as Gabriel and Inamdar (1983) have studied the effects of $CaCl_2$ and organic cationic polymers on the stabilization of clays and fines, the purpose being to restrict migration of these particles in the porous medium which could cause subsequent mechanical blockage of narrow pore throats when exposed to aqueous solutions of lower salinity. Clays also can increase the viscosity of the aqeuous phase and interfere with the diffusion of gases and nutrients essential for microbial metabolism.

The massive amount of surface area (a cubic foot of rock can possess a surface area equivalent to several football fields) contained within a reservoir also can affect

TABLE 3-IV

Percentage distribution of major lithology types [a]

State (number of reservoirs)	Silt-stone	Shale	Chert	Anhy-drite	Igneous rocks	Sand-stone	Lime-stone	Dolo-mite	Carbon-ate	Other
California (875)	1.4	1.6	0	0	0	98.6	0	0	0	0
Oklahoma (5715)	0	0.1	2.0	0	0	74.9	22.1	2.1	0	0
Texas (6197)	0.3	0.5	1.4	0.3	0	62.1	25.6	13.8	0	0.3
Colorado (732)	0.5	2.5	0	0.3	0.1	94.5	3.7	2.5	0	0.1
Louisiana (2052)	0	0	0	0.3	0.1	92.0	8.2	0	0	0.3
Kansas (1526)	0	0.6	4.5	0	0	23.4	49.9	26.0	0	1.0
Mississippi (924)	0.3	0.7	0	0.1	0	93.6	5.5	1.1	0.1	0.6
New Mexico (875)	0	1.4	0	4.7	0	39.8	18.5	35.0	0	0.5
Wyoming (990)	0.2	1.4	0	0	0	84.3	6.2	7.7	0.1	0.1

[a] Information in Tables 3-IV to 3-XV pertaining to reservoir parameters was obtained in data base files ERDA and Oily as compiled by the Office of Information Systems Programs (University of Oklahoma, Norman, Okla.). These data represent a compilation of the information gathered by D.M. Munnecke and J.B. Clark under contract DOE/BC/10169-10 (1979). The numbers in parentheses following each state represent the number of reservoirs sampled by the data base.

microbial growth and activity. It has been known for some time that surfaces can concentrate nutrients and allow bacteria to grow at very low concentrations of nutrients (ZoBell and Grant, 1943). Surfaces also allow for the attachment and colonization of microbial forms capable of forming biofilms on these surfaces, thereby reducing subsequent liquid flow through the rock matrix. Kepkay and Nealson (1982) have shown that increased surface area enhanced the ability of a marine *Bacillus* to sporulate as well as to oxidize manganese. Conversely, Speier and Malek (1982) have recently shown that surfaces can exhibit cytotoxic effects. They demonstrated that surfaces such as silica can adsorb cationic chemicals, which when added in low concentrations would render the surfaces toxic to bacteria on contact. It is not clear, however, if the mechanism of cytotoxicity is due to an increase in effective concentration of the chemical or due to activation or modification imposed by the surface. For a fine discussion of adhesion of microorganisms to surfaces the reader is referred to a book by Bitton and Marshall (1980) as well as a review by Daniels (1972).

In MEOR operations, it may be necessary to transport cells as well as nutrients into the reservoir; therefore, the effect of rocks and clays on the retention of cells and nutrients is of paramount concern. Jang et al. (1983) have shown that synthetic microspheres (1.0 μm in diameter), as well as spores of a *Clostridium* sp., are adsorbed less when transported through sandstone that has been previously oil-saturated. They also found that vegetative cells of several bacterial types when injected into brine-saturated Cleveland Sandstone were strongly adsorbed and subsequently formed a filter-cake at the inlet surface of the core. This also was observed by Jenneman et al. (1983,1984) using a *Pseudomonas* isolate injected into Berea Sandstone. Jang et al. (1983) further implicated the possible use of chemical additives such as Tween 80 or chelating agents to modify the adsorptive capacity of the rock and clay surfaces so that cells may be transported more freely. Jenneman et al. (1984) have shown that nutrients such as proteins and phosphates are strongly adsorbed by Berea Sandstone, whereas glucose and ammonia nitrogen are readily transported through Berea cores within the first several pore volumes of injected fluid.

Porosity and permeability

Oil, along with gas and water, are trapped within void spaces referred to as pores. These pores comprise the void volume of the rock which can be expressed as a fraction or percentage of the bulk volume, which is called porosity, ϕ:

$$\phi = (\text{pore volume}/\text{bulk volume}) \times 100\%$$

Obviously, only intercommunicating pores and corresponding effective porosity (ϕ_e) should be considered.

TABLE 3-V

Percentage distribution of average porosity of oil-bearing strata [a]

State (number of reservoirs)	Percentage of reservoirs with porosity (%) of:					Porosity (%)		
	0.1–10.9	11.0–17.9	18.0–25.9	26–39.9	⩾ 40.0	avg.	max.	min.
California (514)	0.8	5.3	41.4	51.2	1.4	26.5	52	2.0
Oklahoma (346)	8.1	56.6	32.5	2.8	0.0	–	–	–
Texas (4445)	19.5	38.7	28.6	15.7	0.1	–	–	–
Colorado (410)	11.9	44.4	41.5	1.7	0.0	16.2	39	0.16
Louisiana (324)	0.9	7.5	26.2	65.4	0.0	–	–	–
Kansas (100)	26.6	59.5	11.4	2.5	0.0	–	–	–
Mississippi (1011)	3.8	23.3	24.2	48.2	0.2	26.5	–	1.4
New Mexico (475)	61.3	26.9	11.6	0	0.0	10.5	25.4	1.5
Wyoming (531)	17.7	56.5	23.9	1.9	0.0	15.0	40	2.5

[a] See comments in Table 3-IV.

–: not available.

The liquid permeability of a porous material is the ability of the material to conduct a fluid. The units of measurement are given in darcys (D), named after Henri Darcy, who developed the following relationship known as Darcy's Law:

$$\frac{q}{A} = \frac{k}{\mu}\frac{\Delta p}{L}$$

where q = flow rate, $cm^3\ s^{-1}$; A = cross sectional area, cm^2; L = length, cm; μ = viscosity, cP; Δp = differential pressure, atm; and k = permeability, D.

Tables 3-V and 3-VI givc values for average porosities and permeabilities that are found in oil reservoirs for nine states in the U.S.A. Except for Oklahoma, New Mexico and Wyoming, at least two-thirds of the average permeability values of the reservoirs were above 100 mD. This trend is also reflected in the porosity data as these same three states have a high percentage of their average porosity values below 18.0%, especially New Mexico, where 61.3% of average porosity measurements lie between 0.1 and 10.9%. A wide range of minimum and maximum values for both permeability and porosity, however, is evident in most states indicating that a high degree of variability exists within each state. It should be pointed out that average permeability values do not reflect the heterogeneity of a reservoir; therefore, a highly stratified reservoir would exhibit a large range of values around the reported average.

Fatt and Davis (1952) have shown that an increase in overburden pressure (compaction pressure) resulted in a decrease in permeability for several sandstones examined. The permeability for eight of the cores tested when placed under an overburden pressure of 3000 psig ranged from 59 to 89% of the permeability when measured under no overburden pressure. However, above a certain compaction pressure (3,000 psig) further pressure increases do not produce large changes in the

TABLE 3-VI

Distribution of average permeability in oil-bearing strata (%) [a]

State	Percentage of reservoirs with permeability (mD) between:						Permeability (mD)		
	1–25	26–75	76–100	101–125	126–500	⩾ 500	avg.	max.	min.
California (240)	5.4	12.9	7.9	4.6	32.5	36.7	616	5,000	1
Oklahoma (183)	40.3	29.6	10.0	3.2	13.4	3.7	–	–	–
Texas (1552)	26.2	23.5	4.9	2.8	24.5	18.2	–	–	–
Colorado (173)	20.4	15.6	15.0	5.4	35.3	8.4	190	2,240	0.15
Louisiana (128)	1.0	5.4	4.3	5.7	51.8	32.0	–	–	–
Kansas (37)	19.0	42.3	3.2	3.2	22.6	9.7	–	–	–
Mississippi (449)	6.0	14.0	6.5	5.8	43.9	23.8	390	6,750	1
New Mexico (269)	63.0	20.9	5.1	2.8	7.1	1.2	47	1,035	0.1
Wyoming (213)	38.0	21.1	12.7	5.2	18.3	4.7	125	2,000	1

[a] See Table 3-IV for comments.

–: not available.

permeability. Collins (1976) reported that for the Woodbine Sandstone a positive correlation existed between increasing average porosity and increasing mean permeability; however, this correlation may not hold true for all sandstones. Burdine et al. (1950) developed a method of relating the median pore size distribution of reservoir rock to the permeability. By this method, the pore throat size distribution is determined from capillary pressure measurements, which through the use of two equations can be used to estimate the permeability. They found that this method of determining permeability correlated well with experimental measurements of permeability; however, there were some exceptions. In general, cores of equal porosity but differing permeability will have pore size frequencies skewed toward the smaller pore sizes for lower permeabilities and toward the larger pore sizes for higher permeabilities.

The pore size is of concern because bacteria which are found in many different morphologies (e.g., rods, cocci, curved rods, tetrads, chains, etc.) can have dimensions of length approximating 0.5–10.0 μm and widths of 0.5–2.0 μm. This means that pore sizes of less than 0.5 μm will place severe restrictions on the ability of these bacteria to be transported through a rock matrix. Updegraff (1983) has stated that pores must be at least twice the diameter of cocci or short bacilli for effective transport to occur. Kalish et al. (1964) using Berea Sandstone of different permeability ranges (high: 278–400 mD; medium: 130–162 mD; and low: 17.7–48.3 mD) found that the median pore size distribution for cores was 3.5–4.0 μm for the low-permeability range, 4.5–5.0 μm for the medium-permeability range, and 5.5–6.0 μm for the high-permeability cores. They also found, however, that the pore size range with the greatest frequency in each permeability class consisted of pore sizes below 0.5 μm. Forbes (1980) has reported that reservoirs with permeabilities of 400 and 100 mD have average pore throat diameters approximating 15 μm and 7 μm, respectively. Permeabilities of 75–100 mD are thought to be the lower limit for effective microbial transport, although reports have been made on the transport of bacteria through cores of less than 75 mD (Hart et al., 1960; Kalish et al., 1964), and in one case penetration through a core of less than 0.1 mD (Myers and McCready, 1966). Raleigh (1962), as well as Hart et al. (1960), found that bacterial penetration and plugging were independent of permeability. Raleigh (1962) reported that bacterial penetration under non-growth conditions could be directly related to the pore geometry factor G, which is a measure of interconnected pore volume and pore size distribution as calculated from capillary pressure and fractional porosity. Therefore, it is still not clear what the relationship is (if one exists) among microbial transport, permeability and porosity. Still, permeability and porosity measurements represent the most readily available petrophysical parameters that may be important for the study of cell transport in a reservoir.

Not only do permeability, porosity, and pore size influence transfer of microbes, but they also can influence growth and metabolism. Zvyagintsev (1970), using capillaries of rectangular cross-section, has shown that pore size can affect growth rate and size of bacterial cells. He found that cells of *Staphylococcus aureus* and *Saccharomyces vini* increased from 350 to 3560 cells and from 400 to 2750 cells, respectively, in 24 hours while growing in capillaries of 400 × 150 μm (thickness × width, ID). When placed in capillaries of 5 × 3 μm in size, no increases in cell number was reported after 24 hours. Furthermore, the average size of the yeast cells (*S. vini*) was reduced from 6.11 μm in the large capillaries to 4.25 μm in the smaller capillaries. Nazarenko et al. (1974) found that the growth rates of the methane-oxidizing bacterium *Methylosinus trichosporium* did not significantly decrease as the thickness of the capillary decreased from 500 to 20 μm; however, the final number of cells in each capillary did decrease significantly with decreasing capillary thickness. It was also observed that the lag phase of growth for this bacterium increased from 48 hours in the 500-μm capillary to over 72 hours in the 20-μm capillary. All observations were made at the gas–liquid interface of the capillary slides so that diffusion of gas was not limiting. Furthermore, Zvyagintsev and Pitryuk (1973) found that decreasing capillary size from 500 × 250 μm to 8 × 4 μm significantly decreased the cell reproduction rate of several bacterial isolates tested (*Serratia marcescens, Staphylococcus aureus, Bacillus subtilus*, and *Rhodotorula glutinis*). These results were confirmed in capillaries where a continuous flow of nutrients was applied to the capillaries. For *B. subtilis*, no increase in cell number was observed in 8 × 4 μm capillaries both under flow and non-flow conditions. Therefore, the absence of growth could not be attributed to a lack of available nutrients, but instead it was speculated that the physical state of the growth medium (activity, viscosity, etc.) was being altered.

Depth

The depth of petroleum reservoirs varies greatly within each state examined. A reservoir in California, U.S.A., had a reported depth of 3 ft, whereas a reservoir in Mississippi reported a total depth of 19,915 ft. Mississippi and Louisiana had greater than 30% of their reservoirs reported at depths greater than 10,000 ft, whereas greater than 90% of the reservoirs in Kansas were reported to be at depths less than 5000 ft. There have been wells drilled in Oklahoma, Texas, Louisiana, and other states to depths greater than 20,000 ft. Depth values are given in Table 3-VII. The depth itself is not limiting to microbial growth, but its effect on temperature and pressure of the reservoir can affect microbial growth and metabolism.

Temperature

Temperature will generally increase with increasing depth; therefore, the depth of the reservoir can be used to estimate the reservoir temperature. The downhole temperature of a formation can be calculated based on the following

TABLE 3-VII

Percentage distribution of oil reservoirs according to depth [a]

State (number of reservoirs)	Percentage of reservoirs with depth (ft):						Depth (ft)		
	⩽ 2001	2001–4000	4001–5000	5001–7500	7501–10,000	⩾ 10,000	avg.	max.	min.
California (1101)	14.2	26.6	13.4	25.2	14.3	6.4	5102	17,892	3
Oklahoma (7286)	14.3	30.5	12.9	26.3	11.4	4.6	–	–	–
Texas (37606)	5.7	19.1	13.6	34.1	20.1	7.3	–	–	–
Colorado (853)	5.0	11.5	27.3	50.1	6.0	0.1	5143	10,074	9
Louisiana (4154)	2.5	8.2	3.9	14.9	26.9	43.4	–	–	–
Kansas (6792)	5.5	64.0	24.3	6.1	0	0	–	–	–
Mississippi (1020)	0.9	3.3	8.8	34.6	16.7	35.7	8545	19,915	540
New Mexico (1079)	10.7	17.1	10.3	16.6	19.8	25.5	6922	16,497	90
Wyoming (1080)	12.8	23.3	11.1	29.1	16.3	7.4	5445	18,200	15

[a] See Table 3-IV for comments.

–: not available.

TABLE 3-VIII
Reservoir temperature distribution [a]

State (number of reservoirs)	Percentage of reservoirs with temperatures (°C):					Temperature (°C)		
	≤ 35	35–45	45–55	55–75	≥ 75°C	avg.	max.	min.
Oklahoma (24)	0	36.0	44.0	16.0	4.0	49.5	78.9	35.6
Colorado (183)	1	1.9	19.0	49.5	30.0	80.6	340.0	31.1
Texas (549)	9.1	14.3	19.0	39.5	18.1	61.4	158.9	−3.9
Mississippi (904)	6	1.0	5.0	35.0	58.0	89.4	210.0	−5.6
New Mexico (203)	8	15.0	8.7	53.0	22.5	64.4	117.8	27.2
Wyoming (256)	13	10.0	20.0	34.0	36.0	79.4	404.4	12.8

[a] Data for California, Kansas, and Louisiana were not sufficient for adequate representation. See Table 3-IV for additional comments.

temperature–depth relationship (Hilchie, 1982):

$$T_f = T_0 + \frac{g_G D_f}{100}$$

where T_f = temperature of any formation, °F; T_0 = mean surface temperature, °F; g_G = geothermal gradient, °F/100 ft; and D_f = depth of the formation, ft.

Geothermal gradients are commonly 1–2°F/100 ft and surface temperatures vary according to the geographical location. The following are a list of some mean surface temperatures (Hilchie, 1982): Oklahoma, > 60°F; Colorado (oil country), 50°F; Wyoming, 40°F; California, 60°F; Alberta, 35°F.

By examining the data for distribution of temperatures for reservoirs in various states (Table 3-VIII), it is apparent that temperatures vary greatly. A maximum temperature of 404.4°C was reported for a Wyoming reservoir to temperatures below freezing reported in Texas and Mississippi *. The temperature of reservoirs is not available for a large percentage of reservoirs found in California, Kansas and Louisiana. Oklahoma had only 24 reservoirs with available temperature data. If the downhole temperature is approximated according to the depth data, however, it is found that greater than 90% of the reservoirs in all states examined (except Mississippi, Louisiana, and New Mexico) have temperatures of less than 185–210°F (85–99°C) using 10,000 ft as the limiting depth (at this depth temperatures begin to approximate 80–100°C).

Some of the most interesting and informative research regarding microbial life at high temperatures has come from studies of thermal springs and fumaroles found in Yellowstone National Park, Iceland, New Zealand and elsewhere (Brock, 1978). Stanier et al. (1970) described thermophiles as microorganisms that have their optimal growth rate between 55° and 75°C. Bacteria and fungi can be isolated which grow within this temperature range; however, the maximal growth tempera-

* It is probable that these sub-zero readings for reservoirs are reporting errors.

ture for fungi is generally below 60 °C. Microbial thermophily is not discussed here because there are already available some excellent, recently published books and reviews (Ameluxen and Murdock, 1978; Brock, 1978; Ljungdahl, 1979). Instead, a few generalizations that might be of particular interest with regards to life at elevated temperatures found in oil reservoirs are presented here.

As previously noted, oil reservoirs can possess temperatures much greater than 100 °C, because the overburden pressure can maintain water in a liquid state at these extreme temperatures. It is now well accepted that bacteria can be found growing in waters that are at or near the boiling point temperature. Some of these organisms have displayed an ability to withstand extremely low pH (less than 2.0) at these high temperatures (Brock et al., 1972). A particularly interesting ecosystem, which has temperatures and pressures in the range of oil reservoirs, are the deep-sea hydrothermal springs on the Galapagos Rift (Corliss et al., 1979; Karl et al., 1980; Jannasch and Wirsen, 1981). Baross et al. (1982) have reported temperatures for these vents of 350 °C at 261.5 atm. Baross and Deming (1983) discovered bacteria in these hydrothermal vents capable of chemolithotrophic growth at temperatures as high as 250 °C with doubling times of 40 minutes. This ecosystem, therefore, should provide some interesting studies for organisms growing at high temperatures and pressures in the future. However, most of the microorganisms isolated to date near these hydrothermal vents have been microorganisms with optimal growth temperatures below 45 °C (Yayanos and Dietz, 1982). Leigh and Jones (1983) reported the isolation of a methanogen (anaerobic methane-producing organism) from these vents which grows optimally at 86 °C, and Baross et al. (1982) have isolated mixed cultures of thermophilic bacteria from the Galapagos vents that only grow at temperatures above 70–75 °C. Based on the assumption that life at high temperatures is limited by the availability of water in the liquid state rather than by the temperature itself, it is possible that oil reservoirs will present an ecosystem in which the upper temperature limit to life for microorganisms may be extended. Evidence of this will have to wait until more effective sampling techniques can be developed for examining these deep reservoirs, especially their sessile populations.

Of particular interest to MEOR processes are those thermophiles that can grow anaerobically. Clostridia are spore-forming obligate anaerobes which ferment organic matter (sugars, cellulose, etc.) to form alcohols, organic acids and gases (CO_2, H_2). Some species of Clostridia are thermophilic (McClung, 1935; Fontaine et al., 1942; Mercer and Vaughn, 1951). Recently, Grula et al. (1983) isolated Clostridia species capable of growth at temperatures of approximately 45 °C, but found that their ability to produce solvents and gases was markedly reduced at high NaCl concentrations (5% w/v). Anaerobic sulfate-reducing organisms have been isolated capable of growth above 50 °C (*Desulfovibrio thermophilus*) (Rozanova and Khudyakova, 1974); however, these organisms are undesirable for MEOR because they can cause corrosion and non-selective plugging of oil formation through the production of water-insoluble sulfides. Methane-producing organisms have also been isolated that can grow at temperatures in excess of 50 °C; they include *Methanobacterium thermoautotrophicum* (Zeikus and Wolfe, 1972). This organism converts carbon dioxide and hydrogen to methane in order to gain energy for metabolism. Recently Zillig et al. (1981) isolated an extremely thermoacidophilic anaerobe from solfataras in Iceland. This organism, *Thermoproteus tenax*, is capable of growth at temperatures in excess of 90 °C on carbohydrate substrates; however, some forms are known to produce H_2S.

Pressure

Pressure gradients range from 0.43 to 1.0 psig/ft depending on the geographical area; however, in some areas the rate of pressure change increases with increasing depth (Amyx et al., 1960). If pressure is estimated according to the depth of the reservoir it can be seen that states such as Mississippi, Louisiana, and New Mexico contain the largest percentage of deep reservoirs and, therefore, would be expected to have reservoirs with the greatest percentage of high pressures. Twenty-five percent or more of the reservoirs in these states were reported to have depths greater than 10,000 ft, which means that pressures in these reservoirs can be expected to be as high or higher than 10,000 psig (680 atm). For most states examined, however, over 90% of the reservoirs would have pressures well below this value. Even in Mississippi and New Mexico the average depth of reported reservoirs is less than 9000 ft.

Extreme pressures are known to have pronounced effects on the growth and metabolism of microorganisms. Excellent reviews on the effects of pressure on microorganisms are those by ZoBell (1970), Morita (1972) and Marquis and Matsumura (1978). Marquis (1983) presented a paper which discussed pressure effects of deep oil formations. Pressures lower than 100–200 atm generally do not have a pronounced effect on microbial metabolism. However, pressures of 500–600 atm are generally considered to be limiting to growth for most known bacteria. The ocean floor presents the best studied ecosystem which possesses high pressures. Kinne (1972) reported the average pressure on the ocean floor to be 380 atm with a maximum of 1160 atm. Only a couple of reports have been made of truly barophilic (organisms that require high pressure for growth) microorganisms (ZoBell and Morita, 1957; Yayanos et al., 1981). Yayanos et al. (1981) isolated a barophilic organism from the Mariana Trench which is capable of growing at pressures exceeding 1000 atm. It was reported, however, that growth at these pressures occurred very slowly with a generation time of 33 hours at 2 °C. For most MEOR processes, barophilic organisms will not be necessary; instead, barotolerant microbes (organisms which can grow at high pressures, but do not require these high pressures for optimal growth) will suffice.

It has been shown that pressure tolerance is dependent on the biophysical conditions present (Marquis and Matsumura, 1978). The ability to grow at high pressures can be shown to depend on the energy source present, inorganic salts present, Eh, pH, and temperature. However, it is difficult in most cases to determine exactly which of these parameters is affecting growth, as they are all affected by pressure. Of particular interest with regards to oil reservoirs is that salts such as NaCl, as well as divalent cations such as Mg^{2+} and Ca^{2+}, which are commonly found in petroleum reservoirs, can confer a greater barotolerance to some marine organisms (Albright and Henigman, 1971; Marquis and Zobell, 1971; Landau and Pope, 1980). Marquis (1976) found that barotolerance is greatest at a temperature slightly above the optimum growth temperature for some mesophiles and a thermophile (*Bacillus stearothermophilus*). Bubela (1983) reported that an increase in pressure from atmospheric to 20,000 kPa (1 atm = 101.3 kPa) increased the maximum temperature for growth and decreased the mean generation time of a rod-shaped anaerobic isolate from 50°C and 17 hours to 65°C and 12 hours, respectively. Furthermore, the shape of this organism changed from a rod to a coccoidal form with the increase in pressure. Marquis and Bender (1979) have recently discovered that they could successfully isolate a variant of *Streptococcus faecalis* (normal variant tolerates a pressure of up to 750 atm) which grows at pressures as high as 1000 atmospheres by gradually increasing pressure of the growth medium after each successive subculture. Adaptation of cultures to higher pressures, therefore, is a possibility.

Dissolved solids

The aqueous phase probably represents the most important part of the reservoir for MEOR because it is in this phase that nutrients for microbial growth and metabolism will be transported. The water found in an oil reservoir is most frequently marine in origin and has been trapped in the rock formation from the geologic past (connate water). However, during secondary and tertiary recovery operations (e.g., waterflooding, and chemical flooding) fresh water is at times injected into an oil reservoir. These fresh waters can then become the source of soluble organics and colloidal particles (e.g., microorganisms) found in petroleum reservoirs. In most cases, it is not desirable or practical to use fresh water for MEOR processes, because it is not available in the massive amounts required for MEOR, and the use of large quantities can jeopardize the available freshwater supply needed for human consumption and agriculture. However, in some chemical flooding operations where the connate or injection waters are highly mineralized and thus non-compatible with the injected chemical, a pre-flush of fresh water to partially dilute the brine may offer the only available alternative. It is important, therefore, that the chemical makeup of reservoir brines be understood so that compatible organisms and nutrients will be selected for the desired MEOR process.

A number of elements and chemicals which are found in reservoir brines along with the average, maximum, and minimum concentrations reported are listed in

TABLE 3-IX

Chemical composition of reservoir brine, sea water and fresh water [b]

Element or chemical	Sample size	Oil reservoir brine (mg/l)			Sea water [a] (mg/l)	Fresh water [a] (mg/l)
		avg.	max.	min.		
Acetate	1,483	350.2	9,867	0	–	–
Aluminum	128	125.4	1,029	0	0.001	–
Barium	5,796	41.9	3,740	0	0.02	–
Bicarbonate	33,629	422.4	13,926	0	27.0	–
Boron	844	29.9	598	0	4.5	–
Bromide	529	376.9	2,903	0	67.0	–
Calcium	35,383	5,764.5	73,832	0	410.0	–
Carbon dioxide	101	413.4	4,772	2.5	–	–
Carbonate	18,582	54.6	198,571	0	28.0	–
Cesium	98	0.3	247,505	0	0.0003	–
Chloride	35,391	64,940.3	248	0	19,000.0	–
Copper	78	2.9	108	0	0.003	0.01
Ferric iron	453	101.0	3,528	0	0.003	0.67
Ferrous iron	637	50.4	8,800	0	0.003	0.67
Fluoride	8	2.7	7	0	1.3	–
Hydrogen sulfide	7,679	36.3	160,265	0	–	–
Hydroxide	4,040	5.3	1,807	0	–	–
Iodine	679	16.2	565	0	0.06	–
Lithium	1,386	13.9	1,730	0	0.17	–
Magnesium	35,372	1,293.0	49,935	0	1,300.0	–
Manganese	137	13.2	441	0	0.002	0.012
Nitrate	21	2.1	10	0	0.67	–
Oxygen	6	0.1	0.6	0	6.0	–
Phosphate	10	6.3	13	0	0.09	–
Rubidium	101	1.6	64	0	0.012	–
Silicon	576	141.7	8,208	0	3.0	–
Sodium	31,896	33,347.5	147	0	11,000.0	–
Strontium	2,471	349.5	10,000	0	8.0	–
Sulfate	33,346	857.1	14,939	0	900.0	–
Sulfide	126	592.0	71,680	0	–	–
Zinc	29	12.8	136	0	0.01	0.01

[a] See Table 3-IV for additional comments.

[b] Anonymous (1971).

–: not determined.

Table 3-IX. In all cases, there is a wide range of concentrations reported for each chemical. It is clear from this table that oilfield brines are on the average more highly mineralized than sea water or fresh water with regard to all comparable ions with the possible exception of magnesium and sulfate, which are found at approximately equivalent concentrations in both sea water and reservoir brine.

Any MEOR process will require the availability of essential nutrients in order for proper growth and metabolism to take place. All microorganisms have major

TABLE 3-X

Average percentage composition of brine in various states [a]

Element or chemical	California	Colorado	Mississippi	New Mexico	Wyoming	Oklahoma	Texas	Louisiana	Kansas
Total solids	2.2	2.3	14.5	8.5	1.3	15.6	9.7	10.1	9.9
Chloride	1.23	1.18	8.8	5.0	0.57	9.4	5.2	5.8	6.0
Sodium	0.78	0.75	4.5	2.4	0.42	4.7	2.6	3.3	3.1
Potassium/sodium	0.52	0.44	4.0	2.7	0.31	4.1	3.7	3.7	2.5
Calcium	0.06	0.09	0.90	0.40	0.05	1.0	0.50	0.30	0.50
Sulfur	–	–	–	–	–	0.04	–	–	–
Magnesium	0.02	0.02	0.11	0.15	0.01	0.18	0.12	0.06	0.16
Phosphate	4.0 [b]	–	–	–	2.0 [b]	8.0 [b]	–	–	–
Nitrate	3.0 [b]	–	–	3.0 [b]	–	–	2.0 [b]	–	–
Iron	0.001	0.006	–	0.005	0.006	0.02	0.007	0.008	0.006
Oxygen	2.5 [b]	–	–	–	–	–	0.2 [b]	–	–
Ammonium	–	–	–	–	–	0.01 [c]	0.01 [c]	0.01 [c]	0.01 [c]

[a] See Table 3-IV for additional comments.

[b] 10^{-4}.

[c] Average for 4 states.

–: not available.

requirements for carbon, nitrogen, phosphate, sulfur, iron, magnesium, potassium, and calcium, whereas some microorganisms have minor requirements (less than 10^{-4} M) for elements such as molybdenum, manganese, zinc, sodium, chloride, selenium, cobalt, copper, nickel, and tungsten. In addition, for the cell to produce energy, an electron acceptor such as oxygen, nitrate, or sulfate must be present for all respiring organisms, whereas a fermentable substrate such as sucrose, glucose, or lactose must be present if fermenting organisms are used. Also, for certain microorganisms, an organic nitrogen source (e.g., protein) is needed along with small amounts of vitamins. It is unlikely, however, that organisms with specific nutrient requirements for amino acids, vitamins, etc., will be useful for MEOR, because these nutrients would be too expensive to inject in large amounts needed for EOR.

From Tables 3-VIII, 3-IX and 3-X it is evident that oxygen will only be found in low concentrations; therefore, anaerobic, energy-yielding pathways will most likely be utilized. Thus, an easily fermentable sugar such as sucrose with or without nitrates will be needed for effective metabolism. In most cases, these chemicals will have to be injected into the formation since they are not found in sufficient quantities in situ. The presence or addition of sulfates is not desirable because this would promote growth of sulfate-reducing organisms which produce H_2S. Hydrogen sulfide gas is very toxic to life forms as well as corrosive to oilfield equipment and can combine with metals to form insoluble complexes that can cause non-selective plugging of the oil-producing formation and equipment. Sulfides have had a long history of causing damage to oil reservoirs and equipment. Therefore, conditions favoring sulfate reduction should be avoided.

It is apparent from Table 3-X that besides sugars and nitrates, nutrients such as phosphate and nitrogen (reduced form: NH_4^+) might be growth-limiting in the reservoir unless they are injected. Phosphates might be expected to be in limited supply in highly mineralized waters because they complex readily with divalent cations such as magnesium or calcium and precipitate. High average concentrations of calcium and magnesium can be found in some reservoirs, especially in the states of Mississippi and Oklahoma (Table 3-X). Phosphate is utilized by the cell for chemical energy production and for synthesis of nucleic acids and phospholipids. Post (1977) reported that in one hypersaline environment (Great Salt Lake) phosphate concentrations are not that low (630–4000 μg/l), but much of this is in particulate form, raising a question as to the availability of this phosphate for microbial metabolism. It is important that the chemical composition of the brine being utilized be known so that the proper amounts of each of these nutrients can be added to the brine. The compatibility of these nutrients with the brine and clays present must be understood, so that effective transport of nutrients takes place without swelling and migration of clays. The likely source of these nutrients will be inexpensive renewable wastes, such as feedstock molasses and sugar beet molasses and, possibly, cheap inorganic materials.

It is also possible that many of the chemicals and elements that appear in the brine can be toxic to microorganisms if they are in concentrations that are too high. Sorensen et al. (1980) have shown that 0.3 mM H_2S affects denitrification of several denitrifying bacteria by inhibiting the reduction of nitric and nitrous oxide. This is significant because denitrification has been suggested as a possible means for metabolism of microorganisms under anaerobic conditions found in oil reservoirs. In general, for heavy metals, concentrations much in excess of 10^{-3} M can become toxic to many microorganisms, whereas high concentrations of light metal cations can have both inhibitory as well as stimulatory effects. Kugelman and Chin (1971) presented an excellent paper describing the toxicity, synergism, and antagonism of light and heavy metals in an anaerobic environment. The determination of metal toxicity in situ is further complicated by the fact that pH, salinity, temperature, and pressure can all affect the solubilization of these metals. Bubela (1983) reported than an increase in temperature from 53° to 63°C rendered copper more toxic to the thermophile *B. stearothermophilis*, and furthermore the presence of magnesium did not alleviate this toxicity as was found at lower temperatures. Capone et al. (1983) have shown that certain heavy metals at concentrations of 1000 mg/l sometimes found in anoxic salt marsh sediments can have a profound effect on carbon cycling between methanogens and sulfate respirers. Discussion of toxicity of light and heavy metals can be found in articles by Hutner (1972) as well as Ehrlich (1978).

Salinity

Average sodium chloride concentrations are very high in brines found in Oklahoma and Mississippi (> 10% w/v) and are relatively low in states such as California, Colorado, and Wyoming. Sodium chloride makes up almost 90% or more of the average total dissolved solids found in reservoirs of each state examined. Therefore, tolerance of microorganisms to sodium chloride is one of the most important characteristics needed for organisms used in MEOR. Bacteria are known that can grow in brines saturated with sodium chloride (approximately 32% w/v). However, little is known about the physiology of halophilic (salt-loving) forms except that they require a minimum of between 10 and 12% (w/v) sodium chloride in order to grow, and also have complex nutritional requirements. Because of these specific growth requirements and the fact that little is presently known about anaerobic halophiles, their role in MEOR processes is unclear; however, several recent reports have revealed the isolation of halophilic methanogens (anaerobic methane-producing bacteria), one of which can grow optimally at NaCl concentrations of 15% (w/v) (Belyaev et al., 1983; Paterek and Smith, 1983; Yu and Hungate, 1983).

The more likely candidates for MEOR would be those salt-tolerant forms which can grow over a wide range of salt concentrations, sometimes referred to as moderate halophiles. Kushner (1978) has listed isolates that grow within the range of 0–30% (w/v) NaCl. Moderate halophiles that can grow anaerobically at temperatures greater than 50°C would be particularly attractive for use in MEOR. Boyer et al. (1973) have reported the isolation of a *Bacillus* sp. (*Bacillus alcalophilus* subsp. *halodurans*) which grows fermentatively in salt concentrations of 7–15% NaCl at temperatures as high as 55°C and at a pH as high as 10. If NaCl concentrations are lowered to 1–5% NaCl, this organism can then grow at a pH around neutrality. It is also possible to isolate species of *Bacillus* which can grow anaerobically in the presence of nitrates, high NaCl concentration (> 5% w/v), and at high temperatures (50°C). In the author's laboratory many such isolates have been collected capable of growth under these conditions. Also of interest is the fact that several investigators have noticed a positive relationship between the ability of organisms to grow at high salt content and their ability to grow at high temperatures, conditions which are often found in oil reservoirs (Stanley and Morita, 1968; Bilsky and Armstrong, 1973; Novitsky and Kushner, 1975; Keradjopoulos and Halldorf, 1977). Rodriguez-Valera et al. (1980) found in continuous cultures of mixed populations of moderate and extreme halophiles that the moderate halophiles predominated over extreme halophiles at high dilution rates of limiting nutrients and at NaCl concentrations of 20–30% (w/v). This implies that moderate halophiles have the capacity to compete favorably at high salt and limiting nutrient concentrations such as might be encountered in MEOR processes. It should be emphasized, however, that although bacteria can grow at high sodium chloride concentrations it may happen that specific metabolic functions such as gas production, solvent production, exopolymer production, etc. are impaired or lost. For those readers interested in additional information on halophilism there are some excellent review articles available (Gibbons, 1969; Dundas, 1977; Kushner, 1978).

pH

Of the biochemical parameters which affect growth and metabolism of microorganisms, pH appears to be the least extreme in oil reservoirs. Table 3-XI gives the distribution of pH found in oil reservoirs for the nine states examined. The lowest average pH being reported is 6.3 for Mississippi and the highest average pH is 7.9 for Wyoming. Over 90% of those reservoirs examined had a pH between 3.0 and 9.9, but there were also pH values reported above 10.0 and below 3.0. In general, microorganisms grow optimally between a pH of 4.0 and 9.0; however, microorganisms have been isolated which are capable of growth at pH values as low as 1.0 and as high as 12.0.

The pH not only can affect growth and metabolism directly but also in an indirect manner by affecting the solubility of toxic materials. Most notable of these effects are the ones that affect the solubilization of heavy metals. Heavy metals can be very toxic to organisms if they are found at levels highly in excess of what is needed for nutrition, generally in the range of 10^{-3}–10^{-4} M. Table 3-IX indicates that heavy metals such as copper, ferric iron, copper, zinc, etc., can exist in concentrations greater than this range; however, organisms are affected in different ways and certainly organisms are known that can tolerate very high concentrations of most any heavy metal (Daniels, 1972).

Although growth of organisms at pH values found in oil reservoirs does not appear to be an insurmountable problem, the control of specific metabolic processes required for some MEOR processes could be complicated by pH gradients or changes in the reservoir. A well-known example is presented by Gottschalk and Bahl (1981) who discussed the effect of pH on production of butanol by *Clostridium acetobutylicum*. At a pH above 5.0, this anaerobic organism, when grown on glucose, ferments the substrate to butyrate, which in turn lowers the pH below 5.0. At a pH below 5.0, butanol begins to form and butyrate production decreases. Other examples of how pH shifts can affect microbial metabolism are discussed by Neijssel and Tempest (1979). For those readers requiring more information about microbial life at extreme pH values, review articles by Langworthy (1978) and Brown et al. (1980), as well as a recent book on alkalophilic microorganisms by Horikoshi and Akiba (1982) are good references.

API gravity

The oleic or oil phase is the third portion of the reservoir that must be considered in MEOR. It is with this phase that the eventual success or failure of MEOR rests. As previously mentioned, Jang et al. (1983) found that the presence of an oleic phase in sandstone cores flooded with bacterial spores resulted in a greater amount of spores transported through the rock than through cores saturated only with brine. The oleic phase, however, can also pose limitations on MEOR. Limitations posed by the oleic phase in MEOR involve either toxicity effects due to the light volatile fractions or the high densities imposed by heavy asphaltic crudes. Fractions of crude oil normally responsible for toxicity to microorganisms are those hydrocarbons with alkane numbers less than ten. Also, the density of the crude oil can pose a problem when contemplating MEOR, because the heavier the crude the more difficult it is to recover using chemical means due to unfavorable mobility ratios between brine and oil. Crude oil densities can be represented as the API gravity (°API):

$$°\mathrm{API} = \frac{141.5}{\mathrm{sp.gr.}_{60°\mathrm{F}/60°\mathrm{F}}} - 131.5$$

where $\mathrm{sp.gr.}_{60°\mathrm{F}/60°\mathrm{F}}$ = specific gravity at 60°F compared to water at 60°F.

API gravity is measured at 60°F and atmospheric pressure. Crude oils with API gravities of less than 18° are generally considered "heavy", becoming solid-like as the API drops below 10°. These types of oil are generally best recoverable through injected steam or in-situ combustion. However, Finnerty and Singer (1983) have recently reported a microorganism that produces a biosurfactant when grown on

TABLE 3-XI
pH distribution [a]

State	Percentage of reservoirs with pH:									Average pH
	⩽ 3.0	3–3.9	4–4.9	5–5.9	6–6.9	7–7.9	8–8.9	9–9.9	⩾ 9.9	
California	0.4	0.1	0.2	1.1	14.4	65.1	17.8	0.5	0.5	7.5
Colorado	0.2	0.1	0.3	0.9	10.8	39.4	42.4	5.0	1.0	7.8
Kansas	1.1	0.9	2.2	6.7	33.4	50.5	4.6	0.4	0.1	6.9
Louisiana	0.8	1.3	5.1	10.9	35.8	39.3	5.7	0.5	0.4	6.7
Mississippi	0	1.0	6.7	29.7	39.7	20.4	1.9	0	0.6	6.3
New Mexico	1.2	0.4	1.0	6.1	29.9	42.4	15.2	3.0	0.8	7.1
Oklahoma	0.9	0.5	4.3	21.8	44.4	22.2	4.6	0.8	0.3	6.4
Texas	0.5	0.8	2.9	8.4	29.8	43.6	11.3	1.5	0.8	7.0
Wyoming	0.1	0.1	0.3	0.7	8.7	39.1	45.3	4.7	0.9	7.9

[a] See Table 3-IV for comments.

TABLE XII

Percentage distribution of API gravity for crude oil in various states [a]

State (number of reservoirs)	Percentage of reservoirs with API gravity (°):				API gravity (°)		
	1–17	17–31	31–42	> 42	avg.	max.	min.
California (990)	15.9	48.9	31.6	3.6	27.2	77	6
Oklahoma (8071)	0.3	6.7	72.1	21.0	–	–	–
Texas (37997)	1.6	9.8	47.0	41.6	–	–	–
Colorado (525)	0.4	8.6	78.3	12.5	38.5	68	16
Louisiana (6313)	0.2	10.6	38.4	50.8	–	–	–
Kansas (9365)	0.3	15.6	76.7	7.5	–	–	–
Mississippi (979)	2.0	12.7	54.6	30.6	39.0	68	9
New Mexico (835)	0.1	9.9	54.0	35.9	41.1	76	16
Wyoming (751)	4.1	31.2	51.1	13.6	33.6	66	11

[a] See Table 3-IV for additional comments.

–: not available.

hexadecane or Cerro Negro PCN-1 crude oil that is capable of reducing the viscosity of this crude oil (8–15° API) from greater than 25,000 cP down to 250 cP. Of the nine states examined, only California was reported to have a large percentage (15.9%) of reservoirs with crude oils below an API gravity of 18°; most states reported crudes with API gravities of greater than 31° (Table 3-XII). In general, the deeper the formation, the higher will be the API gravity.

It is a well-known fact that biodegradation of crude oil can occur under aerobic conditions (Westlake, 1983); however, less is known of the biotransformation and biodegradation of petroleum under anaerobic conditions found in most reservoirs. It is possible that oxygen could be transported into a reservoir via the injection of surface waters. In fact, Ivanov and Belyaev (1983) found that in the Bondyuzskoe oilfield in the U.S.S.R. the addition of fresh water to the reservoir resulted in bacterial oil oxidation at the interface between fresh and stratal waters. This oxidation resulted in the production of low-molecular-weight organics which would then be used to feed the anaerobic methane-producing populations. Moses et al. (1983) reported the isolation of anaerobic enrichments capable of growth on crude oil in the presence of nitrates as well as growth of a methane-producing enrichment of methanogens in the presence of Forties crude oil. These enrichments were reported to be mixed cultures and significant levels of methane (higher than control volumes) were reported to occur after 25 weeks of incubation. Ward and Brock (1978) found that under aerobic conditions increasing salt concentrations reduced the rate of hydrocarbon metabolism in natural samples from the Great Salt Lake. At concentrations above 20% salinity, they were unable to enrich for microorganisms that could utilize mineral oil as a sole carbon and energy source. Furthermore, Schwarz et al. (1975) reported that increasing pressure at ambient temperatures significantly reduces the rate of hydrocarbon metabolism compared to rates under ambient temperatures and atmospheric pressure. Therefore, the question of significant anaerobic metabolism of hydrocarbons, especially under conditions that exist in many oil reservoirs, is still one of some doubt.

Toxic chemicals

Other chemicals with known toxicity to microorganism that might be found in some reservoirs have been listed by Silvestro and Desmarais (1980). These chemicals include cosurfactants, surfactants, biocides, ethylenediaminetetracetate, and toluene, many of which are used in various chemical EOR operations. Grula and Grula (1983) have determined the toxicity of a great many of these chemicals to microbial soil populations as well as selected strains of microorganisms.

Feasibility

Based on the aforementioned discussion of limitations posed on MEOR by the reservoir, it is possible to arrive at a prediction of the percent of reservoirs treatable for the nine states examined. Table 3-XIII, proposed by Clark et al. (1981) indicates the percentage of reservoirs potentially treatable using the following criteria:

NaCl content in formation water:	less than 10% w/v
pH:	between 4 and 9
Permeability:	above 75 mD
API gravity:	greater than 18°
Temperature:	less than 75°C

TABLE XIII
Percent of reservoirs in each state potentially treatable by MEOR [a]

State	Salt, <10%	pH, 4-9	Permeability, > 75 mD	API gravity, >17°	Temperature, < 75°C	Percentage treatable [b]
Oklahoma	53.5 [c]	97.0	31.0	99.0	96.0	15.5
Texas	53.5 [c]	97.0	51.0	98.0	82.0	21.1
Louisiana	53.5 [c]	97.0	93.5	99.0	75.0	36.0
Kansas	53.5 [c]	97.0	39.0	99.0	98.0	19.6
California	100.0	98.5	82.0	84.0	80.0	54.0
Colorado	100.0	94.0	65.0	99.6	70.0	42.0
Mississippi	15.0	98.5	80.0	98.0	41.0	4.7
New Mexico	100.0	94.5	39.0	99.9	78.0	28.7
Wyoming	100.0	94.2	41.0	95.9	64.0	23.0

[a] See Table 3-IV for comments.
[b] Percentage calculated by considering each parameter as an independent variable.
[c] Average for 4 states.

TABLE 3-XIV

Physical characteristics for reservoirs contained in thirty-five of the United State's largest oilfields [a]

Oilfield	pH	Salinity (% w/v)	(°C)	Depth (ft)			Average permeability (mD)	API gravity (°)	Lithology [b]
				min.	avg.	max.			
1. Prudhoe Bay, Alaska	–	–	79	–	–	–	230 ± 40	26 ± 2	S, L
2. East Texas, Texas	–	–		–	3650	–	–	39	S
3. Wilmington, Calif.	7.5	3.0		3	4170	9200	800 ± 800	22 ± 10	S
4. Yates, Texas	–	–	32	1100	1300	1500	18	30 ± 1	S
5. Midway-Sunset, Calif.	7.5	1.7		1200	3140	5300	1580 ± 1500	19 ± 12	S
6. Wasson, Texas	–	–	52	4900	7260	8800	11 ± 9	36 ± 5	D
7. Kern River, Calif.	7.3	0.1		900				13.0	S
8. Sho-Vel-Tum, Okla.	6.7	11.7	45	230	4800	10,000	1830 ± 3200	32 ± 26	S
9. Panhandle, Texas	–	–	35	2000	2850	3200	21 ± 29	30 ± 4	S, L, D
10. Elk Hills, Calif.	7.4	2.7		2900	6300	9300	–	25 ± 16	S
11. Ventura, Calif.	7.6	3.0		2700	8200	12,000	62	30 ± 5	S
12. Huntington Beach, Calif.	7.4	2.8		2100	4200	6800	1080 ± 600	21 ± 7	S
13. Hawkins, Texas	–	–		3400	5100	7300	–	44 ± 25	S
14. Greta-Tom O'Connor, Texas	–	–		1000	4500	7900	420 ± 180	39 ± 20	D, L, S
15. Long Beach, Calif.	–	2.9		4400	7500	10,000	–	28.5 ± 4	–
16. Eunice Area, La.	7.7	2.4		3700	5600	9200	–	26	S
17. Goldsmith-Andector, Texas	–	–	48	2700	6500	9400	21 ± 20	40 ± 6	D, S, L

18. Coalinga, Calif.	–	–		1900	2500	3200	1600 ± 500	19 ± 4	S
19. McElroy-Dune, Texas	–	–	36	2900	6500	11,700	144 ± 70	36 ± 6	D, L, S
20. Rangely, Colo.	7.5	5.0	71	1700	3700	6000	12	38 ± 5	S
21. Hastings, Texas	–	–		2600	4500	6200	900	42 ±15	S
22. Salt Creek, Texas	–	–	53	1700	6100	8200	17	45 ±14	S
23. Oklahoma City, Okla.	5.8	22.0		–	4150	–	90	39 ± 2	S, L
24. Caillou Island, La.	–	–		7600	16,000	21,700	–	37 ± 6	S
25. Conroe, Texas	–	–		950	3800	5200	900	34 ±21	S
26. Bradford, Texas	–	–	54	6000	7300	9600	17	51 ±14	S
27. Buena Vista, Calif.	7.4	5.8		1800	5200	10,000	740	35 ±20	S
28. Webster, Texas	–	–		1100	2200	3400	–	28 ± 5	S, L, D
29. Santa Fe Springs, Calif.	7.7	2.5		3400	6300	9400	20 ± 5	32 ± 4	S
30. Bay Marchand Block #2, La.	–	–		7900	11,300	12,900	–	31 ±15	S
31. Spraberry Trend, Texas	–	–	59	6200	8400	11,000	–	39 ±13	S
32. Van, Texas	–	–		1200	2900	5200	3000	34 ± 6	S
33. Burbank, Okla.	6.6	11.7		2700	2800	3100	10	38	S
34. McArthur River, Alaska	–	–		–	9350	–	73 ± 29	35 ± 2	S
25. Seminole, Okla.	6.7	15.8		2300	4200	11,000	–	32 ± 7	S

–: not available.

[a] See Table 3-IV for additional comments.

[b] S = sandstone, D = dolomite, L = limestone.

If each criterion is considered to be an independent variable, then the percent treatable can be calculated from the product of the individual percentages. Mississippi had the lowest percentage of reservoirs deemed treatable (4.7%) because of the high salinity found in many of the reservoirs, while California represented the most promising state for MEOR with over half of their reported reservoirs being deemed

TABLE XV
Comments on treatability of giant oilfields by MEOR processes

Oilfield	Reserves ($\times 10^6$ barrels)	Comments
1. Prudhoe Field, Alaska	9600	would require thermophilic microbes
2. East Texas, Texas	12960	not enough information
3. Wilmington, Calif.	830	good chance
4. Yates, Texas	1390	permeability is too low
5. Midway-Sunset, Calif.	565	crude is heavy, otherwise good chance
6. Wasson, Texas	808	permeability is too low
7. Kern River, Calif.	941	heavy crude
8. Sho-Vel-Tum, Okla.	391	good chance, high salt
9. Panhandle, Texas	115	permeability is too low
10. Elk Hills, Calif.	1016	excellent chance
11. Ventura, Calif.	210	low permeability, some thermophilic conditions
12. Huntington Beach, Calif.	139	good chance
13. Hawkins, Texas	438	not enough information
14. Greta-Tom O'Connor, Texas	307	excellent chance
15. Long Beach, Calif.	59	good chance if permeability not too low
16. Eunice, La.	245	good chance if permeability not too low
17. Goldsmith-Andector, Texas	181	low permeability
18. Coalinga, Calif.	191	fair chance, heavy crude
19. McElroy-Dune, Texas	326	permeability may be too low
20. Rangely, Colo.	261	low permeability
21. Hastings, Texas	234	excellent chance if salinity and pH are reasonable
22. Salt Creek, Texas	200	low permeability
23. Oklahoma City, Okla.	19	high salt, low permeability
24. Caillou Island, La.	206	high temperatures
25. Conroe, Texas	177	excellent chance if salinity and pH are reasonable
26. Bradford, Texas	19	low permeability
27. Buena Vista, Calif.	31	excellent chance
28. Webster, Texas	201	not enough information
29. Santa Fe, Calif.	23	low permeability
30. Baty Marchand, La.	149	high temperatures
31. Spraberry Trend, Texas	152	not enough information
32. Van, Texas	115	excellent chance if salinity and pH are reasonable
33. Burbank, Okla.	33	low permeability, high salt
34. McArthur River, Alaska	231	low permeability and high temperatures
35. Seminole, Okla.	219	high salt

treatable by MEOR. Almost all states had at least 20% of their reservoirs deemed treatable using the above criteria. The parameters which seemed to be limiting most often for MEOR in the states examined were permeability and temperature.

Giant oilfields

A search was also performed on thirty-five of the United States giant oilfields. Table 3-XIV lists the physical properties reported for reservoirs within these fields. A single oilfield can contain hundreds of different reservoirs so that variability within any single field should be expected. Table 3-XV gives the total reserves of oil remaining in these fields at the time of reporting and gives a brief statement as to their potential for MEOR methods. It appears that many of these giant fields have either a good or excellent chance to be treated by MEOR taking into consideration presently proposed limitations for MEOR. However, further research is necessary to better define these limitations and possibly to discover other limitations not yet foreseeable.

MICROBIAL ENHANCED OIL RECOVERY APPLICATION

In developing microbial enhancement processes, it must be considered that the products formed to aid in oil release are of biological origin. Thus, they will be subject to biological degradation. If the product of cell metabolism is formed in situ, it will be initially protected to some extent by being in a physically controlled ecosystem, which may not contain many types of microorganisms that would be capable of degrading the compound. The production of the compound, however, would create a selective environment that would be selective to those cells eventually capable of carrying out the degradation. One should not conclude, therefore, that any biological product would be saved from breakdown merely because it was in the reservoir. If, however, the cells that produced the bioproduct are present in the reservoir and are carrying out active metabolism, the rate of production should, in many cases, be greater than the rate of decomposition. If the biological product was synthesized above ground and introduced into the reservoir, there would be no biosynthesis of the material in the reservoir to overcome the effects of biodegradation. Thus, an in-situ approach has advantages over an injection approach.

When examining potential mechanisms by which microbiological activity could aid in the release of oil, a good starting place is to see how many of the known tertiary recovery mechanisms could be carried out by such activities. The following mechanisms could be attained by using the proper culture of microorganism:

(1) Biosurfactant production
(2) Biopolymer production
(3) Gas production (such as CO_2)
(4) Solvent production (such as acetone, alcohol, etc.)

(5) Acid production (organic acids which could react with the formation to increase permeability and release CO_2)

(6) Selective plugging (growth of the cells in the areas of highest permeability, where the cell mass would reduce permeability and thus divert subsequent flooding through unswept areas).

To this list must be added another possibility: some unknown mechanism which has not yet been recognized. This could even involve the growth of cells as firmly attached colonies, which would physically displace the adsorbed oil from the rock surface.

As with any tertiary recovery technique, the process must be chosen after consideration of all parameters of the reservoir. All such factors including the environmental factors, which would influence microbial growth, must be considered when evaluating the potential application of a microbial enhancement process. Consideration must also be given as to whether the microbial process is to be only a near-wellbore treatment or is to involve a penetration deep into the reservoir for a more conventional MEOR treatment.

Most of the MEOR field tests conducted to date have been near-wellbore treatments. It can be considered that these processes are merely specialized bore clean-out procedures. However, the results in many cases are greater than that which would be expected from merely cleaning up the wellbore. Examination of the information available in the literature suggests that this type of treatment is, in itself, worth further development. In such processes, the microorganisms chosen would normally be a mixture of cells capable of producing gas along with solvents and/or acids. A high level of metabolic activity would be desirable.

When developing a microbial process for deep penetration and reservoir treatment, a different approach would be used. Due to the long time periods required to inject the cells deep into the reservoir, it would be necessary to use dormant forms, such as spores, which could be activated by nutrients when the spores are in place, or to use cells with low metabolic activity and long generation times, which could be injected along with the required nutrients. Rapid multiplication of cells during the injection process could decrease permeability by plugging of the formation in the area of injection. Either case would require the ability to control the penetration of cells and their adsorption so that they would be located at the proper place and not be displaced by flow of nutrients or other liquids in subsequent injection processes.

Reliable, predictable processes will have to be developed for cell and nutrient injection for both near-wellbore treatment and for deep reservoir treatment. Various laboratories are working on such problems at present, but the scale-up from a laboratory procedure to a field process may involve much more development.

Microorganisms used will need to be chosen for the type of recovery process desired for a particular application and for cells that will grow in the environment of the reservoir to be treated. It will be necessary to accumulate a collection of microorganisms with desirable characteristics. Growth limitations would need to be established for each culture. From such a collection, the best organism for a particular application would be chosen. The alternative would be to genetically engineer the microorganisms needed for any particular application.

CONCLUSIONS

The use of microorganisms in enhanced oil recovery procedures is feasible as a routine process for those cases in which such use would be advantageous. To determine the conditions under which microbial procedures would be preferred will require an accumulation of information resulting from field applications of the processes. More knowledge must be obtained from laboratory approaches, but without carefully planned and controlled field work, the true advantages of using microorganisms will not be evaluated.

The limitations on growth caused by extreme reservoir conditions are not yet fully evaluated. However, a significant number of reservoirs will be treatable, even under a conservative set of environmental limitations. It is obvious that not all reservoirs will be amenable to microbial processes, but no single chemical enhanced oil recovery process can be used under all conditions. It is not realistic to think microbial processes will become a universal procedure for oil recovery. It is reasonable, however, to think that microbial enhancement of oil recovery will become an accepted process to use where indicated, and to be added to the arsenal of procedures available for enhanced recovery. Economic concerns, although justified, are impossible to evaluate given the present state of the art for this newly developing technology.

ACKNOWLEDGEMENTS

I would like to especially thank Dr. J.B. Clark for his valuable comments and contributions to this manuscript. Also, special thanks go to Dr. D.M. Munnecke for his contributions and to Dr. M.J. McInerney for his comments and suggestions. Finally, I express my sincere gratitude to Susan Russell for her patience in preparing this manuscript for publication. Much of the data contained in this manuscript was compiled under Department of Energy Contract DOE/BC/10169-7, as well as under a grant from the Energy Resources Center at the University of Oklahoma.

REFERENCES

Albright, L.J. and Henigman, J.F., 1971. Seawater salts—hydrostatic pressure effects upon cell division of several bacteria. *Can. J. Microbiol.*, 17: 1246–1248.

Ameluxen, R.E. and Murdock, A.L., 1978. Microbial life at high temperatures: mechanisms and molecular aspects. In: D.J. Kushner (Editor), *Microbial Life in Extreme Environments*. Academic Press, New York, N.Y., pp. 218–278.

Amyx, J.W., Bass, D.M. and Whiting, R.L., 1960. *Petroleum Reservoir Engineering*. McGraw-Hill, New York, N.Y., 610 pp.

Anonymous, 1971. *Marine Chemistry. A Report of the Committee on Oceanography*. U.S. National Academy of Washington, D.C.

Baross, J.A., Lilley, M.D. and Gordon, L.I., 1982. Is the CH_4, H_2, and CO venting from submarine hydrothermal systems produced by thermophilic bacteria? *Nature*, 298: 366–368.

Baross, J.A. and Deming, J.W., 1983. Growth of 'black smoker' bacteria at temperatures of at least 250°C. *Nature*, 303: 423–426.

Beckmann, J.W., 1926. Action of bacteria on mineral oil. *Ind. Eng. Chem. News Ed.*, 4: 3.

Belyaev, S.S., Wolkin, R., Kenealy, W.R., DeNiro, M.J., Epstein, S. and Zeikus, J.G., 1983. Methanogenic bacteria from the Bondyuzhskoe Oil Field: general characterization and analysis of stable-carbon isotopic fractionating. *Appl. Environ. Microbiol.*, 45: 691–697.

Bilsky, A.Z. and Armstrong, J.B., 1973. Osmotic reversal of temperature sensitivity in *Escherichia coli*. *J. Bacteriol.*, 113: 76–81.

Bitton, G. and Marshall, K.C., 1980. *Adsorption of Microorganisms to Surfaces*. John Wiley and Sons, New York, N.Y., 439 pp.

Boyer, E.W., Ingle, M.B. and Mercer, G.D., 1973. *Bacillus alcalophilus* subsp. *halodurans* subsp. nov.: an alkaline-amylase-producing alkalophilic organism. *Int. J. Sys. Bacteriol.*, 23: 238–242.

Brierley, C.L., Brierley, J.A., Norris, P.R. and Kelly, D.P., 1980. Metal-tolerant micro-organisms of hot, acid environments. In: G.W. Gould and J.E.L. Corry (Editors), *Microbial Growth and Survival in Extremes of Environment*. Academic Press, London, pp. 37–51.

Brock, T.D., 1978. *Thermophilic Microorganisms and Life at High Temperatures*. Springer-Verlag, New York, N.Y., 465 pp.

Brock, T.D., Brock, K.M., Belley, R.T. and Weiss, R.L., 1972. Sulfolobus: a new genus of sulfur-oxidizing bacteria living at low pH and high temperature. *Arch. Mikrobiol.*, 84: 54–68.

Brown, M.H., Mayes, T. and Lelieveld, H.L.M., 1980. The growth of microbes at low pH values. In: G.E. Gould and J.E.L. Corry (Editors), *Microbial Growth and Survival in Extremes of Environment*. Academic Press, London, pp. 71–98.

Bubela, B., 1983. Combined effects of temperature and other environmental stresses on microbiologically enhanced oil recovery. In: E.C. Donaldson and B.J. Clark (Editors), *Proceedings, 1982 International Conference Microbial Enhancement of Oil Recovery*, NTIS, Springfield, Va., pp. 118–123.

Burdine, N.T., Gournay, L.S. and Reichert, P.P., 1950. Pore size distribution of reservoir rocks. *Trans. AIME*, 189: 195.

Capone, D.G., Reese, D.D. and Kiene, R.P., 1983. Effects of metals on methanogenesis, sulfate reduction, carbon dioxide evolution, and microbial biomass in anoxic salt marsh sediments. *Appl. Environ. Microbiol.*, 45: 1586–1591.

Clark, J.B., Munnecke, D.M. and Jenneman, G.E., 1981. In situ microbial enhancement of oil production. *Dev. Ind. Microbiol.*, 22: 695–701.

Collins, R.E., 1976. *Flow of Fluids through Porous Materials*. The Petroleum Publishing Co., Tulsa, Okla., 270 pp.

Corliss, J.B., Dymond, J., Gordon, L.I., Edmond, J.M., Von Herzen, R.P., Ballard, R.D., Green, K., Williams, D., Brainbridge, A., Crane, K. and van Andel, T.H., 1979. Submarine thermal springs on the Galapagos Rift. *Science*, 203: 1073–1083.

Daniels, S.L., 1972. The adsorption of microorganisms onto surfaces: a review. *Dev. Ind. Microbiol.*, 13: 211–252.

Davis, J.B., 1956. *Petroleum Microbiology*. Elsevier, Amsterdam, 604 pp.

Dundas, I.E.D., 1977. Physiology of halobacteriaceae. *Adv. Microbiol. Physiol.*, 15: 85–120.

Ehrlich, H.L., 1978. How microbes cope with heavy metals, arsenic and antimony in their environment. In: D.J. Kushner (Editor), *Microbial Life in Extreme Environments*. Academic Press, London, pp. 381–408.

Fatt, I. and Davis, D.H., 1952. Reduction in permeability with overburden pressure. *Trans. AIME*, 195: 329.

Finnerty, W.R. and Singer, M.E., 1983. Microbial enhancement of oil recovery. *Biotechnology*, 1: 47–54.

Fontaine, F.E., Peterson, W.H., McCoy, E., Johnson, M.J. and Ritter, G.J., 1942. A new type of glucose fermentation by *Clostridium thermoaceticum* N. sp. *J. Bacteriol.*, 43: 701–715.

Forbes, A.D., 1980. Micro-organisms in oil recovery. In: D.E.F. Harrison, I.J. Higgins and R. Watkinson (Editors), *Hydrocarbons in Biotechnology*. Heyden and Son, London, pp. 169–180.

Gabriel, G.A. and Inamdar, G.R., 1983. An experimental investigation of fines migration in porous media. *Proc. 58th Annu. Tech. Conf. Exhibition*, SPE 12168. 1–7.

Gibbons, N.E., 1969. Isolation, growth, and requirements of halophilic bacteria. In: J.R. Norris and D.W. Ribbons (Editors), *Methods in Microbiology, 3B*: pp. 79–115.

Gottschalk, G. and Bahl, H., 1981. Feasible improvements of the butanol production by *Clostridium acetobutylicum*. In: A. Hollaender (Editor), *Trends in the Biology of Fermentations for Fuels and Chemicals*. Plenum Press, New York, N.Y., pp. 463–471.

Grula, M. and Grula, E.A., 1983. *Biodegradation of Materials Used in Enhanced Oil Recovery*. National Petroleum Council, Washington, D.C., Final Report, DOE/BC/10304-17.

Grula, E.A., Russell, H.H., Bryant, D., Kenaga, M. and Hart, M., 1983. Isolation and screening of Clostridia for possible use in microbially enhanced oil recovery. In: E.C. Donaldson and J.B. Clark (Editors), *Proceedings, 1982 International Conference on Microbial Enhancement of Oil Recovery*. NTIS, Springfield, Va., pp. 43–47.

Hart, R.T., Fekete, T. and Flock, D.L., 1960. The plugging effect of bacteria in sandstone systems. *Can. Min. Metal. Bull.*, July: 495–501.

Hilchie, D.W., 1982. *Applied Openhole Log Interpretation*. Douglas W. Hilchie, Inc., Golden, Colo.

Hitzman, D.O., 1956. Recovery of oil from oil sands and the like. *U.S. Patent*, No. 2,907,389.

Hitzman, D.O., 1962. Microbial secondary recovery. *U.S. Patent*, No. 3,032,472.

Hitzman, D.O., 1965. Use of bacteria in the recovery of petroleum from underground deposits. *U.S. Patent*, No. 3,185,216.

Hitzman, D.O., 1967. Oil recovery process using aqueous microbiological drive fluids. *U.S. Patent*, No. 3,340,930.

Hitzman, D.O., 1983. Petroleum microbiology and the history of its role in enhanced oil recovery. In: E.C. Donaldson and J.B. Clark (Editors), *Proceedings, 1982 International Conference on Microbial Enhancement of Oil Recovery*. NTIS, Springfield, Va., pp. 162–218.

Horikoshi, K. and Akiba. T., 1982. *Alkalophilic Microorganisms — A New Microbial World*. Japan Scientific Press, Tokyo, and Springer-Verlag, Berlin, 213 pp.

Hutner, S.H., 1972. Inorganic nutrition. *Annu. Rev. Microbiol.*, 26: 313–346.

Ivanov, M.V. and Belyaev, S.S., 1983. Microbial activity in waterflooded oil fields and its possible regulation. In: E.C. Donaldson and J.B. Clark (Editors), *Proceedings, 1982 International Conference on Microbial Enhancement of Oil Recovery*. NTIS, Springfield, Va., pp. 48–57.

Jang, L.K., Sharma, M.M., Findley, J.E., Chang, P.W. and Yen, T.F., 1983. An investigation of the transport of bacteria through porous media. In: E.C. Donaldson and J.B. Clark (Editors), *Proceedings, 1982 International Conference on Microbial Enhancement of Oil Recovery*. NTIS, Springfield, Va., pp. 60–70.

Jannasch, H.W. and Wirsen, C.O., 1981. Morphological survey of microbial mats near deep-sea thermal vents. *Appl. Environ. Microbiol.*, 41: 528–538.

Jenneman, G.E., Knapp, R.M., Menzie, D.E., McInerney, M.J., Revus, D.E., Clark, J.B. and Munnecke, D.M., 1983. Transport phenomena and plugging in Berea sandstone using microorganisms. In: E.C. Donaldson, and J.B. Clark (Editors), *Proceedings, 1982 International Conference on Microbial Enhancement of Oil Recovery*. NTIS, Springfield, Va., pp. 71–75.

Jenneman, G.E., Knapp, R.M., McInerney, M.J., Menzie, D.E. and Revus, D.E., 1984. Experimental studies of in-situ microbial enhanced oil recovery. *Soc. Pet. Eng. J.*, 24: 33–37.

Jones, L.W., 11967. Aerobic bacteria in oil recovery. *U.S. Patent*, No. 3,332,487.

Kalish, P.J., Stewart, J.A., Rogers, W.F. and Bennett, E.O., 1964. The effect of bacteria on sandstone permeability. *J. Petrol. Technol.*, 16: 805–814.

Karl, D.M., Wirsen, C.O. and Jannasch, H.W., 1980. Deep-sea primary production of the Galapagos hydrothermal vents. *Science*, 207: 1345–1347.

Kepkay, P.E. and Nealson, K.H., 1982. Surface enhancement of sporulation and manganese oxidation by a marine bacillus. *J. Bacteriol.*, 151: 1022–1026.

Keradjopoulos, D. and Halldorf, A.W., 1977. Thermophilic character of enzymes from extreme halophilic bacteria. *FEMS Microbiol. Lett.*, 1: 179–182.

Kinne, O., 1972. Pressure, general introduction. In: O. Kinne (Editor), *Marine Ecology, 1, Part 3*. Wiley-Interscience, New York, N.Y., pp. 1323–1360.

Kugelman, I.J. and Chin, K.K., 1971. Toxicity, synergism, and antagonism in anaerobic waste treatment processes. *Adv. Chem. Ser.*, 105: 55–89.

Kushner, D.J., 1978. Life in high salt and solute concentrations: Halophilic bacteria. In: D.J. Kushner (Editor), *Microbial Life in Extreme Environments*. Academic Press, London, pp. 317–368.

Landau, J.V. and Pope, D.H., 1980. Recent advances in the area of barotolerant protein synthesis in bacteria and implications concerning barotolerant and barophilic growth. *Adv. Aquatic Microbiol.*, 2: 49–76.

Langworthy, T.A., 1978. Microbial life in extreme pH values. In: D.J. Kushner (Editor), *Microbial Life in Extreme Environments*. Academic Press, London, pp. 279–315.

Leigh, J.A. and Jones, J.W., 1983. A new extremely thermophilic methanogen from a submarine hydrothermal vent. In: *Abstracts of the 83rd Annual Meeting of the American Society for Microbiology, New Orleans, La., March 6–11*, p. 142.

Ljungdahl, L.G., 1979. Physiology of thermophilic bacteria. *Adv. Microbiol. Physiol.*, 19: 149–243.

Marquis, R.E., 1976. High-pressure microbial physiology. *Adv. Microbiol. Physiol.*, 14: 159–241.

Marquis, R.E., 1983. Barobiology of deep oil formations. In: E.C. Donaldson and J.B. Clark (Editors), *Proceedings, 1982 International Conference on Microbial Enhancement of Oil Recovery*. NTIS, Springfield, Va., pp. 124–128.

Marquis, R.E. and Bender, G.R., 1979. Isolation of a variant of *Streptococcus faecalis* with enhanced barotolerance. *Can. J. Microbiol.*, 26: 371–376.

Marquis, R.E. and Matsumura, P., 1978. Microbial life under pressure. In: D.J. Kushner (Editor), *Microbial Life in Extreme Environments*. Academic Press, London, pp. 105–158.

Marquis, R.E. and Zobell, C.E., 1971. Magnesium and calcium ions enhance barotolerance of streptococci. *Arch. Microbiol.*, 79: 80–92.

McClung, L.S., 1935. Studies on anaerobic bacteria, IV. Taxonomy of cultures of a thermophilic species causing "swells" of canned food. *J. Bacteriol.*, 29: 189–203.

Mercer, W.A. and Vaughn, R.H., 1951. The characteristics of some thermophilic, tartrate-fermenting anaerobes. *J. Bacteriol.*, 62: 27–37.

Morita, R.Y., 1972. Pressure: bacteria, fungi and blue-green algae. In: O. Kinne (Editor), *Marine Ecology, 1, Part 3*. Wiley-Interscience, New York, N.Y., pp. 1361–1388.

Moses, V., Robinson, J.P., Springham, D.G., Brown, M.J., Foster, M., Hume, J., May, C.W., McRoberts, T.S. and Weston, A., 1983. Microbial enhancement of oil recovery in North Sea reservoirs: a requirement for anaerobic growth on crude oil. In: E.C. Donaldson and J.B. Clark (Editors), *Proceedings, 1982 International Conference on Microbial Enhancement of Oil Recovery*. NTIS, Springfield, Va., pp. 154–157.

Myers, G.E. and McCready, R.G.L., 1966. Bacteria can penetrate rock. *Can. J. Microbiol.*, 12: 477–484.

Nazarenko, A.V., Nesterov, A.I., Pitryuk, A.P., and Nazarenko, V.M., 1974. Growth of methane-oxidizing bacteria in glass capillaries. *Mikrobiologiya*, 43: 146–151.

Neijssel, O.M. and Tempest, D.W., 1979. The physiology of metabolite overproduction. *Symp. Soc. Gen. Microbiol.*, 29: 53–82.

Novitsky, T.J. and Kushner, D.J., 1975. Influences of temperature and salt concentration on the growth of a facultative halophilic *Micrococcus* sp. *Can. J. Microbiol.*, 21: 107–110.

Paterek, J.R. and Smith, P.H., 1983. Isolation of a halophilic methanogenic bacterium from the sediments of Great Salt Lake and a San Francisco Bay saltern. In: *Abstracts of the 83rd Annual Meeting of the American Society for Microbiology, New Orleans, La., March 6–11*, p. 140.

Post, F.J., 1977. The microbial ecology of the Great Salt Lake. *Microbial Ecol.*, 3: 143–165.

Raleigh, J.T., 1962. *The Effect of Rock Properties on Bacteria Plugging in Reservoir Rocks.* M.S. Thesis, University of Alberta, Edmonton, Alta.

Rodriguez-Valera, F., Ruiz-Berraquero, F., and Ramos-Cormenzana, A., 1980. Behaviour of mixed populations of halophilic bacteria in continuous culture. *Can. J. Microbiol.*, 26: 1259–1263.

Rozanova, E.P. and Khudyakova, A.I., 1974. A new nonspore-forming thermophilic sulfate-reducing organism, *Desulfovibrio thermophilus* Nov. sp. *Mikrobiologiya*, 43: 1069–1075.

Schwarz, J.R., Walker, J.D. and Colwell, R.R., 1975. Deep-sea bacteria: growth and utilization of *n*-hexadecane at "in-situ" temperature and pressure. *Can. J. Microbiol.*, 21: 682–687.

Silvestro, E. and Desmarais, A., 1980. *Toxicity of Chemical Compounds Used for Enhanced Oil Recovery.* Final Report, DOE/BC/10014-5, pp. 2-1 and 2-2.

Sorensen, J., Tiedje, J.M. and Firestone, R.B., 1980. Inhibition by sulfide of nitric and nitrous oxide reduction by denitrifying *Pseudomonas fluorescences. Appl. Environ. Microbiol.*, 39: 105–108.

Speier, J.L. and Malek, J.R., 1982. Destruction of microorganisms by contact with solid surfaces. *J. Colloid Interface Sci.*, 89: 68–76.

Stanier, R.Y., Doudoroff, M. and Adelberg, E.A., 1970. *The Microbial World*. Prentice-Hall, Englewood Cliffs, N.J., 316 pp.

Stanley, S.O. and Morita, R.Y., 1968. Salinity effect on the maximal growth temperature of some bacteria isolated from marine environments. *J. Bacteriol.*, 95: 169–173.

Updegraff, D.M., 1983. Plugging and penetration of petroleum reservoir rock by microorganisms. In: E.C. Donaldson and J.B. Clark (Editors), *Proceedings, 1982 International Conference on Microbial Enhancement of Oil Recovery*. NTIS, Springfield, Va., pp. 80–85.

Updegraff, D.M. and Wren, G.B., 1953. Secondary recovery of petroleum oil by *Desulfovibrio. U.S. Patent*, No. 2,660,550.

Veley, C.D., 1969. How hydrolyzable metal ions react with clays to control formation water sensitivity. *J. Petrol. Technol.*, 21: 1111–1118.

Ward, D.M. and Brock, T.D., 1978. Hydrocarbon biodegradation in hypersaline environments. *Appl. Environ. Microbiol.*, 35: 353–359

Westlake, D.W.S., 1983. Microbial activities and changes in the chemical and physical properties of oil. In: E.C. Donaldson and J.B. Clark (Editors), *Proceedings, 1982 International Conference on Microbial Enhancement of Oil Recovery*. NTIS, Springfield, Va., pp. 102–111.

Yarbrough, H.F. and Coty, V.F., 1983. Microbially enhanced oil recovery from the Upper Cretaceous Nacatoch Formation, Union County, Arkansas. In: E.C. Donaldson and J.B. Clark (Editors), *Proceedings, 1982 International Conference on Microbial Enhancement of Oil Recovery*. NTIS, Springfield, Va., pp. 149–153.

Yayanos, A.A. and Dietz, A.S., 1982. Thermal inactivation of a deep-sea barophilic bacterium, isolate CNPT-3. *Appl. Environ. Microbiol.*, 43: 1481–1489.

Yayanos, A.A., Dietz, A.S. and Von Boxtel, R., 1981. Obligately barophilic bacterium from the Mariana Trench. *P.N.A.S. USA*, 78: 5212–5215.

Yu, I.K. and Hungate, R.E., 1983. Isolation and characterization of an obligately halophilic methanogenic bacterium. *Abstr., 83rd Ann. Meet. Am. Soc. Microbiol., New Orleans, La., Mar. 6–11*, p. 139.

Zeikus, J.G. and Wolfe, R.S., 1972. *Methanobacterium thermoautotrophicum* sp.n., an anaerobic, autotrophic, extreme thermophile. *J. Bacteriol.*, 109: 707–713.

Zillig, W., Stetter, K.O., Schafer, W., Janekovic, D., Wunderl, S., Holz, I. and Palm, P., 1981.

Thermoproteales: a novel type of extremely thermoacidophilic anaerobic Archaebacteria isolated from Icelandic solfataras. *Zentralbl. Bakteriol., Parasitenkol., Infektionskr. Hyg., Abt. 1: Orig.*, C2: 205–227.

ZoBell, C.E., 1946. Bacteriological process for treatment of fluid-bearing earth formation. *U.S. Patent*, No. 2,413,278.

ZoBell, C.E., 1947a. Bacterial release of oil from oil-bearing materials, 2. *World Oil*, 126: 35–41.

ZoBell, C.E., 1947b. Bacterial release of oil from sedimentary materials. *Oil Gas J.*, 446: 62–65.

ZoBell, C.E., 1947c. Bacterial release of oil from oil-bearing materials. *World Oil*, 126: 36–47.

ZoBell, C.E., 1953. Recovery of hydrocarbons. *U.S. Patent*, No. 2,641,566.

ZoBell, C.E., 1970. Pressure effects on morphology and life processes of bacteria. In: A.M. Zimmerman (Editor), *High Pressure Effects on Cellular Processes*. Academic Press, New York, N.Y., pp. 85–130.

ZoBell, C.E. and Grant, C.W., 1943. Bacterial utilization of low concentrations of organic matter. *J. Bacteriol.*, 45: 555–564.

ZoBell, C.E. and Morita, R.Y., 1957. Barophilic bacteria in some deep-sea sediments. *J. Bacteriol.*, 73: 563–568.

Zvyagintsev, D.G., 1970. Growth of microorganisms in thin capillaries and films. *Mikrobiologiya*, 39: 161–165.

Zvyagintsev, D.G. and Pitryuk, A.P., 1973. Growth of microorganisms in capillaries of various sizes under continuous flow and static conditions. *Mikrobiologiya*, 42: 60–64.

Chapter 4

GEOBIOLOGY AND MICROBIOLOGICALLY ENHANCED OIL RECOVERY

B. BUBELA *

INTRODUCTION

Microbiologically enhanced oil recovery (MEOR) utilizes microorganisms and their products to improve the recovery of crude oil from reservoir rocks. The microorganisms produce metabolites which are introduced into the reservoir to decrease interfacial tension in the oil–water–rock system, to improve mobility ratio of the reservoir fluids, or to induce preferential plugging of pore spaces to facilitate a better directional distribution of sweeping fluids. The selection of the appropriate organisms and their activities are independent of the environmental conditions in the reservoir under consideration. An alternative approach is to produce metabolites with the above properties within the reservoir itself. The selection of organisms and their application in situ is then dictated by the environmental conditions present in a given reservoir. The implications involved in the in-situ MEOR are discussed in this chapter.

GEOBIOLOGY

It is easily understood that when microbiological activity takes place in a geological environment such as a reservoir, the geological, mineralogical, hydrological, and geochemical components of such an environment will have a profound effect on the microorganisms, their biological activity, and their production of the required metabolite. Similarly, the environment, i.e., the reservoir rocks and the reservoir fluids, will be affected to some extent by the biological activity and the biological products. Geobiology involves the study of such an interaction between the geological and biological components of the environment. In broad terms, such an interaction may involve any form of life, but in this chapter only microorganisms and, more generally, geomicrobiology are considered.

---

* Deceased.

Involvement of geobiology in MEOR

The whole concept of microbiologically enhanced oil recovery can be divided into three parts:

(1) Geology and mineralogy

(2) Fluids

(3) Biology

Any significant beneficial changes due to biological activity introduced into the geological and mineralogical component are limited, with the exception of some minor diagenetic changes in the mineral component as discussed later. One may consider this component, therefore, as being non-manipulative.

Fluids, the second component of the system, can be subject to manipulation. Changes in the relative viscosities of the oil–water system (changes in mobility ratio), changes in the interfacial tension between water and oil, the composition of the aqueous phase and, to some extent, the composition of the oil phase can be biologically altered.

The third component, biology, is the manipulator, through which the beneficial changes in the fluids may be achieved.

After defining the manipulator and the components to be manipulated, it is possible to specify what changes in the fluid system would be the most beneficial from the point of enhanced oil recovery in a given reservoir.

IN-SITU MICROBIAL ENHANCED OIL RECOVERY

To be able to consider the implication of geobiology in MEOR, general principles and problems associated with the recovery processes have to be considered.

Enhanced oil recovery techniques are generally applied to reservoirs where production rates are low due to the exhaustion of primary drive responsible for the flow of crude oil or where oil is distributed in such a form that its recovery requires significant changes in parameters governing its flow rates. These parameters are usually interfacial tension of the water–oil–rock system, mobility ratios of the aqueous phase and the crude oil, additional drive, and permeability. A number of abiological techniques have been considered and some tested. They have been summarized by Fayers (1981). Lievens (1983) stated that the methods now being developed and used for enhanced oil recovery are techniques which have been studied for more than two decades. Their development is essentially evolutionary with low probability of an accelerated progress. It is necessary, therefore, to introduce a new generation of processes. Higher-risk research of new and innovative processes has to be initiated. The three most promising candidates are:

(1) Steamflooding of light oil reservoirs

(2) Oil mining

(3) Microbiologically enhanced oil recovery

The following discussion concentrates on the microbiological techniques used in MEOR.

The basic questions to be asked before biotechnology is considered as an alternative method to non-biological techniques are:

(a) Can biotechnology achieve something that abiological methods cannot?

(b) What are the advantages and disadvantages of using biotechnology?

The answer to the first question is a simple NO. Even biotechnology, despite having a mystic ring these days, cannot do anything that is contrary to basic physical and chemical concepts; however, some processes may be facilitated by biotechnology.

From past experience and from the results obtained in laboratories, three major biotechnological approaches may be contemplated:

(a) Significant decrease in the interfacial tension of the water–oil system.

(b) Improvement in the mobility ratio.

(c) Selective plugging.

Before a specific organism can be selected for microbiologically enhanced oil recovery, the environment and the conditions under which it has to operate have to be known.

This knowledge is necessary for the following reasons: Firstly, the biological activity of the organisms and their production of the required metabolite will be influenced by environmental conditions like temperature, pressure, availability of a suitable substrate, electrolyte concentration, and crude oil composition. Secondly, the effectiveness of the metabolite itself will be influenced by some if not all of the parameters just mentioned.

Interfacial tension

The reasons leading to chemically or biologically enhanced oil recovery using surfactants are obvious from the following example:

In water-wetted reservoirs, a substantial amount of the residual oil is located in the form of individual droplets and ganglions. Considering a droplet of oil residing in a pore throat 0.4 mm long with end curvatures of $R_1 = 9 \times 10^{-3}$ mm and $R_2 = 4 \times 10^{-2}$ mm, respectively, the water–oil interfacial tension, σ, being 30 mN m^{-1}, the pressure difference Δp can be determined using Laplace's equation:

$$\Delta p = 2\sigma\left(\frac{1}{R_1} - \frac{1}{R_2}\right) \tag{4-1}$$

The differential Δp required to move this oil droplet through its limiting pore would be 1 – 3 MPa. The practical limits achievable in the field are usually in the range of 20–30 $\times 10^{-3}$ MPa. This example indicates clearly the necessity for an ultra-low interfacial tension of less than 10^{-2} mN m^{-1} to be achieved by the introduction of a surfactant, to obtain a significant enhancement in recovery.

Regardless of the surfactant formulation, to achieve a displacement of a residual oil the following criteria have to be satisfied:

(1) The surfactant employed has to be capable of mobilizing residual oil.

(2) The ability to displace oil must be maintained as the surfactant progresses from the injection point or the location of its biological production to the production well.

(3) An optimal mobility relationship between the crude oil and the aqueous phase must be satisfied.

Since the amount of surfactant produced by an organism is finite in its concentration, consideration must be given to mechanisms rendering it ineffective. The major ones are:

(1) Retention of the surfactant by the porous matrix.

(2) Mixing and dilution of the surfactant with and by the reservoir fluids.

(3) Partitioning of the surfactants between the oil and the aqueous phase.

(4) Biological degradation.

To sustain a favorable mobility ratio, the viscosity of the surfactant solution should be in the vicinity of 8 cP to avoid viscous instabilities that could augment mixing of the surfactant with the reservoir fluids thus decreasing the miscibility (Pasquarelli and Wasan, 1978).

There is probably no single surfactant which would work satisfactorily in all reservoirs. This is simply because the operational situation varies from reservoir to reservoir, depending on the geological, mineralogical, and chemical characteristics of the reservoir, its physical parameters, and the physicochemical composition of the reservoir fluids. This observation is generally true, regardless of the origin of the surfactant, whether biological or abiological. Akstinat (1981) suggested that inasmuch as conditions in a reservoir are very complicated, they should be studied using simulated model systems.

Akstinat (1981) described general conditions present in a number of reservoirs. He observed that many solutes present in reservoir waters are far above the limits indicated by their solubility constants. The increase of solubility is caused by the formation of metal organo-complexes. All samples investigated from about 60 reservoirs contained heavy metals and sulfates in solution. The calcium and magnesium concentrations were high, whereas those of strontium and barium were lower. The reservoir pressures were between 5000 and 10,000 kPa, the temperature range was 40–80°C, CO_2 concentration in the aqueous phase was 4 g per 100 g of water, and the pH ranged from 3.0 to 6.5. Wright (1963) reported that at pressures above 6000 kPa, the pH stabilizes at 3.3 due to the presence of CO_2. All of these environmental conditions are well within the range of biological survival. As the biologically produced surfactant has to be operative under the above-mentioned conditions, a brief discussion of some characteristics of the surfactants may be helpful.

Surfactants

Surfactants can be divided into the following groups: anionic, cationic, amphoteric, and non-ionic. Biological activity can produce surfactants belonging individually to all of these groups:

(1) Anionic: containing carboxylic acids.
(2) Cationic: including amines and heterocycles.
(3) Amphoteric: represented by amino acids and peptides.
(4) Non-ionic: like esters.

The surfactants have different abilities to reduce interfacial tension and characteristics of crude oils depending on the composition (paraffinic, naphthenic, aromatic or mixed), the colloidal chemistry of the crude oil and, thus, on the presence of asphaltenes and resins. The efficiency of the surfactants depends on the environmental conditions in the reservoir like diffusion rates from the bacterial side to the oil–water interfaces. The diffusion coefficient of the surfactant depends on the viscosity of the reservoir fluids. The efficiency of the anionic and amphoteric surfactants is affected by the Ca^{2+} and Mg^{2+} cations in the reservoir waters.

The efficiency of the surfactant is also affected by the lithology of the reservoir rock. Different rocks adsorb surfactant to a different degree. The ability of clays to adsorb surfactants decreases in the following order: cationic, non-ionic, anionic compounds. Silicates show slight adsorption of non-ionic surfactants, but adsorb strongly cationic surfactants. Akstinat (1981), therefore, indicated that cationic surfactants should not be applied to silicate-rich reservoirs. For enhanced oil recovery, a surface adsorption of a surfactant of 0.5×10^{-4} mg cm^{-2} on quartz is generally considered as an acceptable level.

A surfactant to be acceptable to MEOR should demonstrate as many as possible of the following characteristics:

(1) Concentration enrichment at the oil–water interface and, thus, biological production of a surfactant at the oil–water interface. This would be a great advantage.
(2) Ability to permanently lower the interfacial tension below 10^{-2} mN m^{-1}.
(3) Partial solubility in oil.
(4) Capability of stabilizing oil–water emulsions.
(5) Solubility or at least dispersibility in highly saline reservoir waters.
(6) Low adsorption coefficient relative to the reservoir rock.

The adsorption of a mixture of a surfactant and a co-surfactant by the reservoir rock may result in a pseudo-chromatographic effect, which would separate the mixture into individual components and consequently reduce the mixture's efficiency.

The presence of high salinity affects considerably the selection of a surfactant. Only non-ionic surfactants with a high degree of ethoxylation ($n > 10$) are soluble significantly in 10% NaCl brines. If a surfactant is to be produced biologically in a highly saline reservoir, the organisms have to be halophilic, and the surfactant should be preferably non-ionic.

Akstinat (1981) observed that oils with an aromatic character increase their interfacial tension with the aqueous phase as its salinity increases.

It is assumed that the surfactant will have to stay in the reservoir for one to two years for its effect to be fully realized. Such a surfactant has to be reasonably resistant to the biodegrading activity of the endogenous biological population or the activity of other microorganisms should a consortium be introduced for MEOR. In general terms, biostability of compounds having surfactant activity decreases in the following order: phosphates, sulphates, carboxylic derivatives, amines.

Environmental stresses and their limiting effect on MEOR

Before discussing specific concepts of geobiology directly connected with the biotechnological aspects of MEOR, it may be appropriate to identify some of the environmental stresses likely to be present in the majority of reservoirs.

It is likely that the temperature of fluids in the selected reservoirs will be in the 50–100°C region or possibly even higher. Such fluids will be depleted of oxygen, and probably the environment will be strongly reducing, having Eh values less than −100 mV. The pressure may reach thousands of kilopascals, and the concentration of electrolytes in the reservoir waters could be up to several hundred parts per million. The availability of readily assimilatory carbon, phosphorus and nitrogen may be limited, while the pH range may be close to 2 or 11. It is quite natural, therefore, to ask if it is at all possible to expect microorganisms to live under such conditions, and to supply the required product.

Biological activity is known to occur under a wide range of environmental conditions. Some of them are listed in Table 4-I.

The various effects of the individual environmental stresses will be first discussed in general terms and then mentioned when appropriate in further specific cases.

Temperature

As evident from Table 4-I, the temperature range where biological activity is feasible is rather broad and exceeds the stability range of pure water. The lower temperature ranges have no relevance to MEOR. Elevated pressures sustain the fluid temperatures above the boiling point of water. A comprehensive table of thermophiles, organisms capable of living at biologically elevated temperatures, was provided by Tansey and Brock (1978).

pH

Most natural environments have a pH value close to 6–8. Extremely high and low concentrations of hydrogen ions are toxic. Cultures of organisms were isolated from environments having pH of 1 and temperature of 55°C (Mosser et al., 1973).

TABLE 4-I

Range of environmental conditions for biological activity (After Trudinger and Bubela, 1967)

Factor	Range	
	from	to
Temperature:	$-24°C$ (fungi, bacteria)	104°C (sulfate-reducing bacteria under 100,000 kPa hydrostatic pressure)
Hydrostatic pressure:	essentially atmospheric	140,000 kPa (deep-sea bacteria)
pH:	pH 1 (*Thiobacillus oxidans*)	pH 13 (*Plectenome nostocorum*)
Eh:	-450 mV and pH 9.5 (sulfate-reducing bacteria)	$+850$ mV, pH 3 (iron bacteria)
Salinity:	double redistilled water	saturated brines (halophilic bacteria)
Aridity:		48% humidity at 20°C (*Pleurococcus vulgaris*)
Radiation:	essentially 0	6×10^4 Gy (*Micrococcus radiodurant*)
Water activity (a_w):	0.65–0.70 (*Aspergillus glaucus*)	essentially 1

Very little is known of the effect of pH on the formation of H_3O^+, $H_5O_2^+$, $H_7O_3^+$, and $H_9O_4^+$ at various temperatures, and their effect on biological activity. High hydronium or hydroxyl ion concentration plays an important role in solvolysis, ionic state of nutrients, and colloidal properties of microenvironments. At lower pH the solubility of heavy metals and, therefore, their toxicity increases. Some organic acids become protonated and may enter cells with toxic effect. At high pH, some essential co-substrates or inorganic components required for growth may precipitate as carbonates, phosphates or hydroxides and may become a limiting parameter in controlling bacterial growth. The distribution of electric charges on bacterial cells depends on the pH of the microenvironment; therefore, bacterial distribution due to their hydrophobic–hydrophilic properties at the oil–water interfaces may become affected by pH.

Salinity

In the living cell, water serves as a medium of interaction between the populations of molecules. Solutes and the structure of water control all vital processes in the cell. A small change in solute concentration and water activity may result in significant physiological changes. Every solute dissolved in water attracts water molecules and, therefore, reduces its freedom. Water activity, which may be calculated from the ratio of the vapor pressure of pure solvent to the vapor pressure of the solution, decreases with the increase in the solute concentration (the activity of pure water being taken as unity). Water stresses on biological activity were discussed by Brown (1976). It seems that biological activity is prevented by water activity below 0.62 limit.

Heavy metals

Heavy metals are present in reservoir waters at ppm concentration. Despite the fact that quite frequently metals at such concentrations are not toxic to microorganisms, their presence may be detrimental to biological activity (Bubela, 1983a) when other environmental stresses, like temperature or limitation in available molecular oxygen, are present. The responses to heavy metals by microorganisms may take a number of forms. Specific requirement for a supplementary substrate may arise and there may be a change in the mean generation time or in morphology of the organism. In some cases, detoxification processes involving metals operated by the cells are simple, like the reduction of sulfate to sulfide by the sulfate-reducing organisms and subsequent precipitation of the toxic metal. In other cases it is more complex, as described by Ehrlich (1964) for the detoxification of arsenopyrite. In some cases, where heavy metals were introduced into the environment simultaneously with an increase in pressure (Acruri and Ehrlich, 1977), they became strongly inhibitory at concentration well below their normal toxic levels.

Pressure

The effect of pressure on biological activity has been reviewed by Marquis and Matsumura (1978) and Marquis (1983) and is discussed elsewhere in this book.

MICROBIOLOGY OF RESERVOIRS

It is difficult to obtain bacterial samples from reservoirs under sterile conditions. Claims concerning the isolation of an endogenous organism from a reservoir, therefore, have to be considered carefully (Kuznetsov et al., 1963; Davies and Updegraff, 1964).

The origin of microbial populations in reservoirs is probably due to the introduction of the organisms from the surface by reservoir waters. Such a source of the reservoir population is quite feasible despite the long periods of time required for the organisms to travel from their place of origin. The surface water usually contains some components capable of supporting bacterial life. Furthermore the waters may become enriched in such components by passing through strata containing organic matter.

Despite the number of limiting parameters affecting microbiological activity in reservoirs, numerous observations on endogenous reservoir populations have been reported. Greve et al. (1957) reported a number of aerobic and anaerobic organisms present in reservoir waters. A great variety of organisms was reported by Lazar (1983), including the genera *Bacillus, Pseudomonas, Micrococcus, Mycobacterium, Clostridium* and some *Enterobacteriaceae*.

Andreyevskiy (1959) has followed the passage of a bacterial population introduced into the reservoir by surface waters from their source to an area where the underground waters emerged again. Microorganisms were present at all points of collection with a maximum concentration of sulfate reducers at the oil–water contact area. This location coincided with maximum H_2S production, thus suggesting that the organisms used some components of the crude oil for their activity.

The organisms most frequently reported in association with oil deposits are sulfate-reducing bacteria (Spurny and Dostalke, 1956; Greve et al., 1957). As these organisms are of practical importance to the oil industry in general, they are discussed here in some detail.

Biological activity of sulfate reducers and the subsequent production of H_2S results in the souring of oil with a consequent corrosion of plant equipment. Moses and Springham (1982) reported a case where sea water was used for sweeping purposes and the oil became so rich in H_2S that it became unsalable. Recovery on changing back to fresh waterflood took several years.

Sulfate reducers are found in environments which may differ considerably in temperature. Thermophilic organisms (Tansey and Brock, 1978) and mesophiles are quite common among sulfate reducers. A psychrophilic strain of sulfate reducers has been described by Barghoorn and Nichols (1961), but little is known of its properties.

Halophilic sulfate reducers have been isolated from oil reservoirs (Maxina and Rozanova, 1978). Salt tolerance is quite a common characteristic of sulfate reducers (Postgate, 1979). Some strains are barotolerant and some are even barophilic (ZoBell and Morita, 1957). Bubela (unpublished results) observed a 10-fold increase in the production of H_2S by a species of *Desulfotomaculum nigrificans* when the organisms were exposed to a pressure of 37,000 kPa. Cell-free preparation of *D. nigrificans* operating under the same conditions produced 3 times as much of H_2S as obtained with whole cells at ambient pressure. No significant increase in biomass was observed during these experiments. This observation eliminates the possibility of an increase in cell permeability or a decrease in the generation time of the organisms as major causes for the increase in the H_2S production. Another example of barotolerance of *Desulfotomaculum* to pressure was reported by Jannasch et al. (1976). *Desulfovibrio* are usually motile and, therefore, independent to some degree for their spreading through the reservoir of the fluids movements. The redox potential of less than -100 mV is sufficient to permit the activity of sulfate-reducing organisms.

The end product of the sulfate reduction is sulfide ion which has an equilibrium Eh (at 15°C and pH 7) in the region of -320 mV. In such a medium, free oxygen can exist only at very low concentration and only anaerobic respiratory processes should take place. That oxygen does exist in free form, under such conditions, has been shown by Beijerinck (as quoted by Postgate, 1979). Sulfate reducers are not true autotrophs as they require organic material for their growth (Postgate, 1979). Widdel's (1980) observations contradict Beijerinck's findings because some of his isolates were grown autotrophically on CO_2 and H_2.

Comprehensive lists of substances inhibitory to sulfate reducers are given by Postgate (1979), Saleh et al. (1964), and Jack and Thompson (1983). Most of these inhibitors listed probably would be equally effective on the majority of microorganisms encountered in a reservoir. A specific inhibition of sulfate reducers with the exception of MoO_4^{2-} (Peck, 1959), therefore, seems improbable. The effect of some of the inhibitors may be enhanced by additional factors like NaCl or iron salts (Costello et al., 1970).

Air is the cheapest inhibitor of sulfate-reducing organisms, but its practical application is limited. The introduction of sufficient oxygen into a reservoir may induce biological and abiological formation of ferric hydroxide and thus potentially plug the reservoir rock. Furthermore, the sulfate reducers are not killed by oxygen, just inhibited; therefore, reintroduction of anaerobic conditions or formation of anaerobic micro-environments would result in renewed production of H_2S. According to N. Millis (private communication, 1982), carbon monoxide killed an unspecified *Desulfovibrio*. The operative mechanism was apparently the inhibition of the iron-containing cytochrome system. The specificity of such an inhibition should be further investigated, as Tatsuhiko (1958) described a *Desulfovibrio* capable of oxidizing CO to CO_2.

Widdel (1980) observed that sulfate reducers, isolated from acetate, stearate and benzoate enrichments, degraded these substrates completely to CO_2. Isolates from propionate, butyrate and palmitate enrichments ceased their biodegradation at the acetate level. This observation would indicate that some strains have specific requirements; therefore, even if sulfate reducers are commonly present in reservoirs, their biological activity and the formation of metabolic products may differ from reservoir to reservoir.

The specific significance of sulfate reducers to MEOR is discussed later.

It is evident from the examples cited that a variety of organisms are able to grow under reservoir conditions. This observation is of great significance to MEOR, because any introduced microorganism will have to compete with the endogenous bacterial population. As it may be necessary, as discussed later, to introduce an additional substrate to the reservoir in order to facilitate the required biological activity, a possible "bloom" of the endogenous population may create problems.

When considering the MEOR techniques, one may pose the following question: Why the biological activity is apparently low in reservoirs where endogenous bacterial populations have been detected? The reasons may be numerous. Conditions in oil reservoirs are usually anaerobic or at their best just microaerobic. So far, no conclusive evidence has been offered that microorganisms can use hydrocarbons as a substrate under such conditions. Readily available carbon, therefore, may be one of the limiting factors. This aspect is discussed later. Furthermore, phosphorus is usually present at concentrations below 0.05 ppm, thus possibly becoming a limiting factor. Practically no data are available from oilfields on the form and concentration of nitrogen in reservoir waters. Its low concentration may be the reason for the low-density endogenous bacterial population. It is envisaged that both elements may need to be supplied into the reservoir before a satisfactory MEOR activity will be achieved.

BIODEGRADATION OF OIL

Because of their low biological activity at the present time, the effect of the endogenous bacterial population on the composition of the crude oil is probably quantitatively insignificant. The situation could change considerably, should the endogenous population increase significantly, because of an increase in the concentration of a readily available substrate or co-factors.

The biodegradation of crude oil is frequently and mistakenly considered to be identical with the biodegradation of hydrocarbons. Crude oil contains a large number of non-hydrocarbon components and any alteration in their qualitative or quantitative composition may significantly alter the characteristics of the crude oil.

Biodegradation processes taking place in a reservoir are important for a number of reasons, two of them at least being of significance to MEOR:

(1) Alteration of physicochemical characteristics of the crude oil due to biological activity.

(2) Utilization of the crude oil as a substrate for the introduced microbiological population.

The action of microorganisms on crude oil was originally reviewed by ZoBell (1946), who recognized that such activity depends strongly on the chemistry of the compounds in the crude oil and on environmental conditions.

Under aerobic conditions, a variety of microorganisms can alter crude oil (Atlas, 1981) as evident from the application of biotechnology to antipollution techniques employed during spills of crude oil and the use of hydrocarbons as substrates for the production of single-cell proteins. The dependence of hydrocarbonoclastic organisms on oxygen was observed by Crow et al. (1976) who found that bacterial density was up to $100\times$ higher at the surface of water than 10 cm below the surface. Such a vertical microbial distribution could have been caused equally well by the hydrocarbons distribution through the water column due to their lower specific gravity.

The situation becomes more complicated and more confused when biodegradation under anaerobic conditions is considered. It has to be stressed again that crude oil is not composed of hydrocarbons only. It contains a variety of organic compounds having a broad spectrum of functional groups and a considerable amount of compounds containing heteroatoms in their structure. So far there is no convincing evidence that microorganisms can biodegrade hydrocarbons anaerobically. The variation in crude oil composition through a reservoir, frequently attributed to microbiological degradation under anaerobic conditions, could be the result of a fractionation of the precursors of the crude oil during their transport from the place of accumulation prior to the formation of hydrocarbons (Zhang Yi-Gang, 1981). The difficulty in determining whether or not there is an absolute absence of molecular oxygen has been the main obstacle in evaluating critically the reported so called "anaerobic" processes. A lack of proper controls and a lack of indication of the absence of other components in a mixture of hydrocarbons used during the experimental work, are some of the objections to the work reported so far (Mahadevan, 1974).

Current evidence supports the view that anaerobic biodegradation proceeds at negligible rates, if at all (Ward, 1980). Some of the reports of anaerobic degradation have not been confirmed (Traxler and Bernard, 1969). Atlas (1981) concluded that anaerobic degradation of oil is insignificant. He further concluded that hydrocarbons which enter anaerobic environments, such as anoxic sediments, are well preserved and persist over geological times. Suggestions that in a natural ecosystem nitrates and sulfates could serve as electron acceptors during anaerobic respiration, using hydrocarbon substrate, have not been biochemically confirmed.

Wake et al. (1977) excluded most hydrocarbons as substrates on thermodynamic grounds, but noticed that alkenes were thermodynamically feasible substrates.

Reports on bacterial degradation in reservoirs are even more confusing. Karaskiewicz (1974) reported a decrease of some crude oil fractions, especially asphaltenes, when microorganisms were introduced into the injection well. He explained this observation by suggesting that the asphaltenes were biodegraded into paraffins. This report has to be evaluated with caution. When Bubela (1983a) incubated a crude oil under anaerobic conditions with a mixed bacterial population isolated from Melbourne sewage, asphaltenes were precipitated. In addition, a high-molecular-weight material was formed, which co-precipitated with the asphaltenes. Karaskiewicz's results, therefore, could be due to the precipitation of the high-molecular-weight components from the oil in the reservoir. The increase in the lower-molecular-weight constituents of the recovered oil was only apparent. Therefore, a biofractionation and not biodegradation could have been responsible for the increase in paraffins in the recovered product.

Kuznetzov et al. (1963) described an experiment where samples of crude oil and core material were supplemented with ammonium phosphate and incubated under apparently anaerobic conditions. Production of nitrogen, methane, CO_2 and, in some cases, small amounts of H_2 was reported. This experiment does not clearly show that the hydrocarbons were biodegraded and it is quite probable that non-hydrocarbon components from the crude oil were involved.

Oil deposits in northern Alberta (Canada) differ significantly from conventional deposits in their high viscosity and asphaltene content. It has been suggested that they originate from subsurface pooled oils altered by bacterial action producing high-density oils (Bailey et al., 1973). An alternative mechanism has been offered by Zhang Yi-Gang (1981), who suggested that such deposits are formed by biodegradation of the original organic material followed by its diagenetic maturation. Reports by Rubenstein et al. (1977) that bitumens are the result of biodegradation, which renders a number of hydrocarbons, have to be considered carefully. The amount of oxygen in the reservoir waters is present in ppm concentrations. Its availability decreases rapidly with the increase of electrolytes in the water. Any aerobic degradation of significant magnitude would require a quantity of oxygenated water greater by several orders of magnitude than the amount of water available at any time. Furthermore, such degradation would take place only at the oil–water interface. Any significant degradation would require a considerable mixing of the oil phase, a rather difficult process in high-viscosity oils.

There is some evidence that bacteria under anaerobic conditions may alter crude oil (V. Moses, private communication). Bubela (1983a) has shown that a material isolated as asphaltene with a molecular weight of 2000–3000 daltons and possessing carboxylic groups was found in some oils after an exposure to anaerobic bacteria at 60°C for 6 weeks. This high-molecular-weight material did not form in the absence of the organisms. No conclusion could be made so far as to the precursors of this material. It is evident that biological activity may alter the physicochemical properties of the crude oil and, therefore, directly affect processes leading to its recovery. Should a precipitation of asphaltene-like material take place in the reservoir rock, its permeability may be considerably decreased and oil recovery hindered.

EFFECT OF MICROBIOLOGICAL ACTIVITY ON OIL RESERVOIRS

The previous sections discussed some of the aspects of environmental stresses encountered by the biological population in reservoirs and, to some extent, the effect of microbes on the composition of the reservoir fluids. As defined in the opening section, geobiology is concerned with the mutual interaction of biology and geology. Therefore, the biological effects on the reservoir rock, its composition and physical properties, have to be considered.

Mineral matter

Carbonates

The most common changes in mineral matter present in reservoirs are the dissolution and precipitation of mineral phases. In some instances, such processes are closely associated with the diagenesis of organic matter. One of the simple steps, which is of significance to MEOR, is dissolution of carbonates due to the biologically formed CO_2, resulting in an increase in porosity and permeability.

Bubela (1983a) observed that samples of calcite and dolomite exposed to activity of an anaerobic bacterial consortium producing CO_2 and organic acids, increased the permeability of a given sample by an order of magnitude in 6 weeks. Calcite was more readily affected than dolomite and the reaction rate depended on the crystallinity of the sample: the finer the crystallinity the slower the rate of dissolution. Precipitation of carbonates due to biological activity may take place when sulfates are reduced biologically. Usually aragonite is formed and a reduction in permeability results due to cementation of pore spaces. More complex relations between organic matter, biological activity and mineralization occur during dolomitization. Dolomitization is the formation of dolomite, a calcium-magnesium carbonate ($CaMg(CO_3)_2$). Dolomite is formed from a simple calcium carbonate like calcite or magnesium carbonate like nesquehonite (Ferguson et al., 1977) in the presence of calcium and magnesium ions and CO_2. Bubela et al. (1975) indicated that fresh as well as decaying organic matter may influence the alkalinity of the interstitial fluids, thus enhancing the dolomitization processes. When nesquehonite was overlying organic matter buried in sediments, it underwent a number of mineralogical changes leading to the formation of huntite and protohydromagnesite (Ferguson et al., 1977), thus decreasing the permeability of the rock. (Also see Chilingar et al., 1979, on dolomites and dolomitization.)

Iron salts

When both iron and sulphate are present in reservoir waters, cementation and plugging of pore spaces may take place. The sulfate is biologically reduced to sulfide, which reacts with the iron and gelatinous iron sulfide, forming hydrotroilite. In some cases, the precipitation of metal sulfide may be hindered by the presence of organic matter as described by Hallberg et al. (1980).

Silicates

The geobiology of silicates is more complicated. Humic acids formed from biologically degraded organic matter are reputed to be strong agents degrading silicate rocks (Zajic, 1969). Microbiological activity resulting in CO_2 and carbonic acid production will normally precipitate any colloidal silica, thus causing a decrease in the permeability of the rock. *Bacillus extorquens* is reported to be able to degrade mica, nephelines, leucite, and olivine (Silverman, 1979). Because of the slow rate of the decomposition of silicates, the significance of these processes to rock permeability and MEOR is rather dubious.

Summary

The most significant geobiological changes affecting mineral components of the reservoir rock may be summarized as follows:

(1) Precipitation of a mineral phase, usually a carbonate, due to an increase of alkalinity of the interstitial waters. An example is the precipitation of calcium

carbonate during biological reduction of sulfates. The process of carbonate precipitation can be summarized as follows (Trudinger, 1982):

$$2CH_2O + SO_4^{2-} \rightarrow H_2S + 2CO_2 + 2OH^- \tag{4-2}$$

$$Ca^{2+} + CO_2 + 2OH^- \rightarrow CaCO_3 + H_2O \tag{4-3}$$

CH_2O represents biodegradable organic matter and OH^- represents the component responsible for the increase in alkalinity.

(2) Precipication of iron sulfide during the same process can be presented as follows:

$$SO_4^{2-} \rightarrow S^{2-}; \quad S^{2-} + Fe^{2+} \rightarrow FeS \tag{4-4}$$

Changes in porosity and possibly permeability by cementation processes, as described by reactions (4-2) and (4-3) may be caused by two mechanisms:

(1) A given volume of one mineral is dissolved and this is followed by the precipitation of another mineral of the same specific volume. The porosity of the reservoir rock will not change but the permeability may be affected by relocation of the newly formed solids resulting in blocked passages between pores (pore throats) or decreased pore size.

(2) If the dissolved mineral and the freshly precipitated material differ significantly in their specific volumes, like gypsum (0.434) and calcite (0.344), a change in porosity of about 26% should result. The permeability will be affected at the same time.

The situation becomes more complicated if instead of specific volumes, bulk volumes (apparent volumes) of the minerals are considered. In some cases, the final effects as described in (2) may be significantly altered or even reversed. The precipitation of carbonates in interstitial pores will have a dual effect. It will plug or reduce the pore spaces and as carbonates have a higher interfacial tension relative to aqueous phase than silicates, the efficiency of oil recovery may change (Tragus, 1975).

Effect of biological activity on the permeability of the reservoir rock

Permeability of the reservoir rock, one of the most important parameters in any oil recovery technique, may be influenced by biological activity in three ways:

(1) Changes in its mineral components as described in the previous section.

(2) By biomass accumulation.

(3) By altering the composition of organic matter in the pores of the reservoir rock.

The above processes may occur individually or simultaneously.

Biomass accumulation

Due to biological activity, films and plugs of organic matter accumulate in interstitial spaces.

This matter may be formed from (a) living bacterial mass, (b) cellular debris, or (c) exocellular material. Bubela (1983a) has shown that the formation of such a plug is dependent not only on the relative sizes of the interstitial pores and the organism (as suggested by Dostalek, 1954) but also on the shape of the organisms. Rod-shaped and filamentous organisms are more likely to reduce the permeability of the rock by plugging than spheroidal (coccoidal) organisms. It is obvious that pore size distribution in the reservoir rock will be an important factor. The removal or destruction of such a plug by an increase in hydrostatic pressure acting parallel to the flow of the reservoir fluids was more effective in the case of the coccoidal organisms than the rod-shaped and elongated forms of organisms (Bubela, 1983a). Similar observations were made by Jenneman et al. (1983).

The shape of the organism may be affected by environmental stresses, like heavy metals (Bubela, 1970), which may induce a considerable increase in size of the organisms and change their shapes from rods to more elongated forms. In some cases, an increase in pressure and the temperature during the bacterial growth induced morphological changes from rods to coccoidal forms (Bubela, 1983b), thus making the organism more suitable to MEOR.

Anisotropic permeability of sediments caused by bacterial activity (Bubela et al., 1975) may affect not only the rate of flow of reservoir fluids, but the direction of the flow as well. As shown experimentally (Bubela, 1980), when fluids pass through boundaries between layers of significantly different permeabilities, the direction of their flow may be deflected and the distance the fluids have to travel may be increased by several orders of magnitude.

Such a change would result in an increase in the time required for the fluids to reach their expected destination. Due to directional changes of the drive mechanism, fluids may appear at a location quite distant from the one originally expected. Such boundaries in permeabilities are readily produced by biological activity in sediments and porous rocks (Bubela, 1980). Should such a change in the direction of flow of the reservoir fluids be caused by an introduced biological activity, a previously active recovery mechanism may become ineffective.

Sharpley (1961) suggested that the rocks with permeability in the Darcy range should not plug due to injection with fluids containing 10^6 cells/ml, whereas for rocks in the 20 mD range the concentration should be less than 3×10^4 cells/ml. This suggestion should be considered with caution as it is quite obvious that it is not only the magnitude of permeability which is important, but also the pore size, pore size distribution, the shape of the organisms, and the relative sizes of the cells and the pores. Cases of plugging due to bacterial mass have been reported (Allread, 1976), where 2×10^7 cells/ml reduced the permeability of a reservoir rock by 80%.

Fatty acids

Biological activity may increase the porosity and permeability of a carbonate reservoir rock by production of CO_2 and organic acids. The organic acids participate in the dissolution of the carbonate rock. Such acids would then form calcium or magnesium salts. With the increasing carbon chain length of the acids, the solubility of their calcium and magnesium salts decreases rapidly as insoluble soaps are formed. Two effects detrimental to oil recovery result from the precipitation of the salts of long-chain fatty acids: (a) a decrease in permeability due to plugging of pore spaces, and (b) an increase in interfacial tension between the pore walls coated with such a hydrophobic film and the aqueous phase.

Inorganic matter

Plugging problems due to the formation of ferric hydroxide by iron bacteria in the presence of oxygen (Allread, 1976) are rather questionable, because of the low redox potential of reservoir waters.

A change in the permeability of an environment due to the introduction of a bacterial population for the purpose of MEOR may be caused in a rather indirect way. In many cases, the reservoir rock contains a significant proportion of a clay material. Such material may be only partially consolidated or be present as free moving particles in the interstitial spaces of the rock (AMDEL, 1982). Clay particles increase in size due to swelling if the salinity of their aqueous environment decreases (Monicard, 1979). Under reservoir conditions the size of the particles is in "equilibrium" with the salt content of the interstitial waters. Should fluids of lesser salinity be introduced into the reservoir, as may be the case during an injection of a bacterial population into the reservoir, or should an attempt be made to propagate microorganisms in such fluids in situ, an increase in size of the clay particles due to their swelling may occur. This will result in a decrease in permeability.

Mechanisms of plugging

Plugging of pore spaces or pore throats may be caused by several mechanisms. The simple passive process consists of bacterial cells being strained by the porous medium from the passing fluids, thus blocking the passage required for the flow of reservoir fluids. The plug produced in such a way would be composed of whole cells, cellular debris, and other particulate material caught in the original bacterial matrix. The establishment of such a plug will depend on the relative sizes of the pores and the organisms, their shapes, particle size distribution, and the morphology of the organism.

Plugs and microbial film may be produced actively by organisms. Some organisms can attach themselves to the surfaces even in cases where they are smaller than the pore throats. The attachment of organisms to solid surfaces depends on the physicochemical characteristics of solids, the characteristics of the fluids carrying the organisms, and the characteristics of the bacterial surfaces. Marshall (1976) ascribed such attachments to structural forms such as pili and biological polymers acting as adhesives. The adhesive properties of organisms are not constant, and may vary under environmental stresses (Brown et al., 1970).

PHYSICOCHEMICAL ASPECTS OF MICROBIAL ECOLOGY AS RELATED TO MICROBIAL ENHANCED OIL RECOVERY

Microbial ecology may be studied on a macroscale, in the case of MEOR, considering the whole reservoir, or on a microscale, where a single organism and a single capillary pore are considered. Environmental variations within several micrometers are significant in relation to the individual organism. A knowledge of the behavior of microorganisms within microhabitats is essential for the definition of their activity in different ecosystems. One of the microhabitats which is important to microorganisms is an interface. An interface may be defined in physicochemical terms as the boundary between two phases in a heterogeneous system.

In the reservoir, interfaces exist between oil and water, fluids and gases, fluids and solids, and solids and gases. The gases may be gaseous hydrocarbons or CO_2. The fluids are crude oil and water. Solid surfaces are provided by rock or the bacterial cell.

Interfaces

Microbial activity near surfaces is affected by microenvironmental changes like pH variation, resulting from the attraction and the repulsion of H^+ ion, concentration of substrates and inhibitors, and the availability of gases.

Microorganisms with hydrophobic surfaces have tendency to accumulate at the water–oil interfaces. Should such organisms produce a surfactant, their location would be of significant advantage to MEOR as it would be produced at the site where it is needed (Rosenberg et al., 1983).

Marshall (1976) suggested that despite their average size being larger than generally accepted for colloidal particles, bacteria in free suspension behave as colloids in an aqueous colloidal suspension. This, Marshall claims, is at least partially due to the mutual electrostatic repulsion between negatively charged bacteria. The charges on bacterial surfaces depend on the pH of their microenvironment. Negative charges are usually due to carboxylic functional groups, which are ionized in alkaline environments with a gradual decrease in ionization at lower pH. Positive charges are due to ionized amino groups at pH below 7. Should the bacterial cell carry both functional groups, its surface charge is difficult to predict and depends on a number of parameters other than pH (Marshall, 1967).

Bacterial cells possessing both hydrophobic and hydrophilic surfaces may behave similar to surfactant molecules in an aqueous suspension. At low concentration, they are singularly distributed through the medium. When the concentration of the cells reaches a critical value, they cluster and form rosettes (Marshall and Cruickshank, 1973) with their hydrophobic parts closely attached inside the rosette, while the hydrophilic parts are facing the aqueous environment.

Under such a condition, microorganisms behave similar to surfactants molecules. Such a situation is called Critical Micelles Concentration (CMC), a parameter important in enhanced oil recovery (Danielsson, 1978).

During the MEOR processes, it is essential that microorganisms remain as "colloidal" suspension to obtain the maximum penetration of the cells into reservoir spaces and not to decrease the permeability of the rock. Suspension of microorganisms is facilitated by repulsive forces due to like electrostatic charges on their surfaces. Low electrolyte concentration in the reservoir waters, therefore, is more favorable to MEOR, as high concentration could in some cases facilitate the formation of bacterial aggregates and diminish the penetrating capacity of the bacterial population.

Microorganisms reach interfaces by Brownian movement, mobility, boyancy, currents in the interstitial fluids, or movement of the surfaces themselves. Mudd and Mudd (1924) observed bacterial streaming at interfaces of fluids. This may be due to the composition of the interfacial film, local inequality in interfacial tension, often resulting from surface-active substances produced by the organisms, or minute currents arising from mixtures of the phases across and along the boundary line.

Microorganisms prefer surfaces of high interfacial tension (Mudd and Mudd, 1924). When interface between two immiscible liquids (e.g., water and oil) moves, it sweeps bacterial masses ahead into the aqueous phase. With films of low interfacial tension, the phase boundary may be stretched into a peninsular projection, which may break off and leave droplets of water containing bacteria in the organic phase (Marshall, 1976). All or some of the above processes may affect the MEOR by influencing the distribution of microorganisms.

Substrate

One of the parameters to be considered in the assessment of a given reservoir for MEOR is the possibility that the substrate introduced to support the bacterial growth may be adsorbed onto the walls of the pores. Such adsorption will depend on the mineralogy, chemistry and ionic state of the rock surface, and the molecular structure and concentration of the organic matter.

Many natural habitats have low nutritional status. Solid surfaces capable of absorbing organic matter are potential sites of nutrient concentration and, therefore, sites of intensified microbial activities. A paucity in substrate availability through the aqueous phase and its concentration on pore surfaces could result in localized formation of biomass and may, thus, cause a decrease in the reservoir rock permeability. If the substrate is adsorbed, its availability to the organisms will depend on the substrate location relative to the organisms, on the presence or absence of extracellular enzymes, the tendency of the enzymes to be adsorbed, and the configuration of the enzyme and the substrate in the adsorbed state. Some substrate (e.g., carbohydrates) react differently with different minerals of the same group. For example, montmorillonite and not kaolinite adsorbs carbohydrates (Filip, 1973). Up to 25% of dextran adsorbed on montmorillonite was not available to bacterial population (Olness and Clapp, 1972). Montmorillonite affects bacterial utilization of both the carbohydrates and the proteins. It adsorbs pepsin and chymotrypsin, enzymes involved in breaking protein molecules and making them available to further bacterial utilization. The activity of the adsorbed enzymes depends on the orientation of the enzyme's active group (Harter and Stotsky, 1973). Therefore, some substrates usually available to an enzyme may become unavailable if the enzyme or the substrate are adsorbed on solid surfaces.

Adsorption of bacteria and chemical compounds on clays is affected by cations present in the reservoir waters. Sodium saturation of clays increases their adsorption capacity, whereas calcium decreases it (Marshall, 1976). The adsorption is further affected by pH, and the ionic strength of the electrolyte present (Lahau, 1962).

It is evident from the above cited examples that plugging and obstruction of interstitial spaces can be affected by a large number of parameters associated with bacterial population in a reservoir.

CONCLUSIONS

Despite the number of problems that have to be solved before MEOR can be safely and usefully applied to reservoirs, MEOR may be considered as one of the useful techniques to improve the supply of crude oil. The main advantages of MEOR is that no high capital investment is required for its installation, the products and byproducts of MEOR are eventually biodegradable, and the running costs are well below the costs of comparative abiological processes.

It is evident that not all the parameters affecting MEOR are understood. One could summarize by suggesting that the following lines of research should be perused to obtain the maximum benefits which MEOR may offer to the oil industry:

(1) Inhibitory and enhancing factors affecting bacterial growth in reservoirs.

(2) Selection and adaptation including genetic manipulation of organisms capable of producing viscosifiers and surfactants.

(3) Behavior of such metabolites under reservoir conditions.

(4) Interaction between organisms and the mineral phase.

(5) Geological, geochemical, hydrological and biological investigation of reservoirs as relevant to MEOR.

ACKNOWLEDGEMENTS

The author is grateful for suggestions and constructive criticisms by J. Bauld, Pamela Stark, P.A. Trudinger and M.R. Walter. The investigation of MEOR is supported by grants from the National Energy Research Development and Demonstration Council. The Baas Becking Geobiological Laboratory is supported by the Bureau of Mineral Resources, Geology and Geophysics, Commonwealth Scientific and Industrial Research Organization, and Australian Mineral Industries Research Association.

REFERENCES

Acruri, E.J. and Ehrlich, H.L., 1977. Influence of hydrostatic pressure on the effect of heavy metal cations, manganese, copper, cobalt and nickel, on the growth of three deep-sea bacterial isolates. *Appl. Environ. Microbiol.*, 33: 282–288.

Akstinat, M.H., 1981. Surfactants for EOR, processes in high-salinity systems. In: F.J. Fayers (Editor), *Enhanced Oil Recovery*. Elsevier, Amsterdam, pp. 43–62.

Allred, R.C., 1976. The role of anaerobic organisms. In: *The Recovery of Oil. Proceedings, 1976 Engineering Conference*. NSF, Washington, D.C., p. 133.

AMDEL, 1982. Report 2/1/10-GS795/82. Australian Minerals Development Laboratory, Freville, SA.

Andreyevskiy, I.L., 1959. The ways of utilizing oil microbiology in the oil extraction industry. *Tr. VNIGRI*, 131 (4).

Atlas, R.H., 1981. Microbial degradation of petroleum hydrocarbons. *Microb. Rev.*, 45: 180–209.

Bailey, N.J.L., Krouse, H.R., Evens, C.R. and Rogers, M.A., 1973. Degradation of crude oils by waters and bacteria. *Bull. Am. Assoc. Pet. Geol.*, 57: 1276–1290.

Barghoorn, E.S. and Nichols, R.L., 1961 Psychrophiles. *Science*, 134: 190–192.

Brown, A.D., 1976. Microbial water stress. *Bacteriol. Rev.*, 40: 803–846.

Brown, C.M., Elwood, D.C. and Hunter, J.R., 1970. Growth of bacteria at surfaces. *FEMS Microbiol. Lett.*, 1: 163–166.

Bubela, B., 1970. Chemical and microbiological changes in *Bacillus stearothermophilus* induced by copper. *Chem.-Biol. Interact.*, 2: 107–116.

Bubela, B., 1980. Some aspects of interstitial water movements in simulated sedimentary systems. *Aust. Bur. Miner. Resour., Geol. Geophys. Bull*, 5: 257–263.

Bubela, B., 1983a. Combined effect of temperature and other environmental stresses on microbiologically enhanced oil recovery. In: E.C. Donaldson and J.B. Clark (Editors), *Proceedings, 1982 International Conference on Microbial Enhancement of Oil Recovery*. U.S. Department of Energy, NTIS, Springfield, Va., pp. 118–123.

Bubela, B., 1983b. Physical simulation of microbiological enhanced oil recovery. In: J.E. Zajic, D.G. Cooper, T.R. Jack and N. Kosaric (Editors), *Microbial Enhanced Oil Recovery*. PennWell Books, Tulsa, Okla., pp. 1–7.

Bubela, B., Davies, A.J. and Ferguson, J., 1975. Biological and abiological processes in a simulated sedimentary system. *J. Geol. Soc. Austr.*, 32 (2): 135–141.

Chilingar, G.V., Zenger, D.H., Bissell, H.J. and Wolf, K.H., 1979. Dolomites and dolomitization. In: G. Larsen and G.V. Chilingar (Editors), *Diagenesis in Sediments and Sedimentary Rocks*. Elsevier, Amsterdam, pp. 423–536.

Costello, D., King, R.A. and Miller, J.D.A., 1970. Influence of sodium chloride on inhibition of *Desulfovibrio* by a surfactant. *Arch. Microbiol.*, 71: 196–198.

Crow, S.A., Cook, W.L., Ahearn, D.G., and Bourquin, A.W., 1976. Microbial populations in coastal surface slick. In: J.M. Sharpley and A.M. Kaplan (Editors), *Proceedings, III Industrial Biodegradation Symposium*. Applied Science Publishers, London.

Danielsson, I., 1978. Lyotrophic liquid crystals in the structure of biomembranes. *Adv. Chem. Ser.* 2: 13–15.

Davies, J.B. and Updegraff, D.M., 1954. Microbiology in petroleum industry. *Bacteriol. Rev.*, 18: 215.

Dostalek, M., 1954. Propane-oxidizing bacteria. *Cesk. Biol.*, 3: 173–181.

Erhlich, H.L., 1964. Bacterial oxidation of arsenopyrite and anargite. *Econ. Geol.*, 59: 1306–1312.

Fayers, F.J. (Editor), 1981. *Enhanced Oil Recovery*. Elsevier, Amsterdam, 596 pp.

Ferguson, J., Bubela, B. and Davies, P.J., 1977. Simulation of sedimentary ore-forming processes: concentration of Pb and Zn from brines into organic and Fe-bearing carbonate sediments. *Geol. Rundsch.*, 64: 767–782.

Filip, P., 1973. Clay minerals as factors influencing biochemical activity of soil microorganisms. *Foll. Microbiol.*, 18: 56–74.

Greve, W., Muller, A. and Swartz, W., 1957. Mikrobiologische Untersuchungen in Verbindung mit Flutung von Erdöllagerstatten. *Zentralbl. Bakteriol., Parasitenkd., Infektionskr. Hyg., Abt.* 2, 110: 82–106.

Hallberg, R.O., Bubela, B. and Ferguson, J., 1980. Simulation of metal chelating in two reducing systems. *J. Geomicrobiol.* 2: 99–113.

Harter, R.D. and Stotsky, G., 1973. X-ray diffraction, electron microscopy, electrophoretic mobility and pH of some stable protein complexes. *Soil Sci. Soc. Am., Proc.*, 37: 116–123.

Jack, T.R. and Thompson, B.G., 1983. Patents employing microorganisms in oil production. In: J.E. Zajic, D.G. Cooper, T.R. Jack and N. Kosaric (Editors), *Microbial Enhanced Oil Recovery*. PennWell Books, Tulsa, Okla.

Jannasch, H.W., Wirsen, C.O. and Taylor, C.D., 1976. Undercompressed microbial populations from the deep sea. *Appl. Environ. Microbiol.*, 32: 360–367.

Jenneman, G.E., Knapp, R.M., Menzie, D.E., McInerney, M.J., Revus, D.E., Clark, J.B. and Munnecke, D.M., 1983. Transport phenomena and plugging in Berea sandstones using microorganisms. In: E.C. Donaldson and J.B. Clark (Editors), *Proceedings, 1982 Industrial Conf. on Microbial Enhancement of Oil Recovery*. U.S. Department of Energy, NTIS, Springfield, Va., pp. 71–75.

Karaskiewicz, J.S., 1974. *Zastosowanie Metod Mikrobiologieznych w Intesifychii Exploatacji Karpackich Zloz Ropy Naftowej*. "Slask", Katovice.

Kuznetzov, S.T., Ivanov, M.V. and Lyalikova, N.N., 1963. *Introduction to Geological Microbiology*. McGraw-Hill, New York, N.Y., 252 pp.

Lahau, M., 1962. Adsorption of sodium bentonite particles on *Bacillus subtilis*. *Plant Soil*, 17: 191–208.

Lazar, I., 1983. Microbial enhancement of oil recovery in Romania. In: E.C. Donaldson and J.B. Clark (Editors), *Proceedings, 1982 International Conference on Microbial Enhancement of Oil Recovery*. U.S. Department of Energy, NTIS, Springfield, Va., pp. 140–148.

Lievens, E.J., 1983. Abstract. In: E.C. Donaldson and J.B. Clark (Editors), *Proceedings, 1982 International Conference on Microbial Enhancement of Oil Recovery*. U.S. Department of Energy, NTIS, Springfield, Va., p. 6.

Mahadevan, A., 1974. Anaerobic degradation of alkenes by microorganisms. *J. Sci. Ind. Res.*, 33: 39–43.

Marquis, R.E., 1983. Barobiology of deep oil formations. In: E.C. Donaldson and J.B. Clark (Editors), *Proceedings, 1982 International Conference on Microbial Enhancement of Oil Recovery*. U.S. Department of Energy, NTIS, Springfield, Va., pp. 124–128.

Marquis, R.E. and Matsumura, P., 1978. Microbial life under pressure. In: D.J. Kushner (Editor), *Microbial Life at Extreme Environments*. Academic Press, New York, N.Y., pp. 105–158.

Marshall, K.C., 1967. Electrophoretic properties of yeast and slow growing species of *Rhizobium*. *Aust. J. Biol. Sci.*, 20: 429–438.

Marshall, K.C., 1976. *Interfaces in Microbial Ecology*. Harvard University Press, Cambridge, Mass., 131 pp.

Marshall, K.C. and Cruickshank, R.H., 1973. Cell surface hydrophobiety and the orientation of certain bacteria at interfaces. *Arch. Microbiol.*, 91: 29–40.

Maxina, T.N. and Rozanova, E.P., 1978. Halophilic organisms S12 from oil reservoirs. *Mikrobiologiya*, 47: 142–148.

Monicard, R.P., 1979. *Properties of Reservoir Rocks and Core Analysis*. Inst. Fr. Pét.

Moses, V. and Springham, D.G., 1982. *Bacteria and the Enhancement of Oil Recovery*. Applied Science Publishers, London, 178 pp.

Mosser, J.L., Mosser, A.G. and Brock, T.D., 1973. Bacterial origin of sulfuric acid in geothermal habitats. *Science*, 179: 1323–1324.

Mudd, S. and Mudd, E.B.H., 1924. The penetration of bacteria through capillary spaces: kinetic mechanisms at interfaces. *J. Exp. Med.*, 40: 633–645.

Olness, A. and Clapp, C.G., 1972. Microbial degradation of montmorillonite-dextran complex. *Soil Sci. Soc. Am., Proc.,* 37: 37–45.

Pasquarelli, C.H. and Wasan, D.T., 1978. Surface phenomena in EOR. In: D. Shah (Editor), *Proceedings, EOR Symposium. August 20–25, 1978, Stockholm*. Plenum, New York, N.Y.

Peck, A.D., 1959. The ATP dependent reduction of sulphate with hydrogen in extracts of *D. desulfuricans*. Proc. Natl. Acad. Sci. U.S., 45: 701–708.

Postgate, J.R., 1979. *Sulphate Reducing Bacteria*. Cambridge University Press, London, 120 pp.

Rosenberg, M., Gutnick, D.L. and Rosenberg, E., 1983. Bacterial adherance to hydrocarbons. In: J.E. Zajic, D.G. Cooper, T.R. Jack and N. Kosaric (Editors), *Microbial Enhanced Oil Recovery*. PenWell Books, Tulsa, Okla., pp. 114–123.

Rubenstein, T., Strauz, O.P., Spyckerelle, C., Crawford, R.J. and Westlake, D.W.S., 1977. The origin of the oil bitumens of Alberta, a chemical and microbiological simulation studies. *Geochim. Cosmochim. Acta,* 41: 1341–1353.

Saleh, A., McPherson, R. and Miller, J.D.A., 1964. The effect of inhibitors on sulphate reducing bacteria. *J. Appl. Bacteriol.,* 27: 281–293.

Sharpley, J.M., 1961. Bacteria in flood waters. *Pet. Eng.*, 33: 55–67.

Silverman, M.P., 1979. Biological and organic decomposition of silicates. *Stud. Environ. Sci.*, pp. 445–465.

Spurny, M. and Dostalek, M., 1956. Mikrobiologie Naftovych Poli. *Pr. Ustavu Naft. Vyzk.*, 59: 26–30.

Tansey, M.R. and Brock, T.D., 1978. Microbial life at high temperatures: ecological aspects. In: D.J. Kushner (Editor), *Microbial Life in Extreme Environments*. Academic Press, New York, N.Y., pp. 159–216.

Tatsuhiko, Y., 1958. Enzymic oxidation of carbon monoxide. *Biochem. Biophys. Acta*, 30: 194–195.

Tragus, F.J., 1975. Adsorption of mixed surfactant systems. *52th Annu. Fall. Meet., SPE-AIME, March 28-April 1, 1975*, SPE 5564.

Traxler, R.W. and Bernard, J.M., 1969. The utilization of *n*-alkanes by *Pseudomonas aeruginose* under conditions of anaerobiosis. *Int. Biodeterior. Bull.*, 5: 21–25.

Trudinger, P.A., 1982. Geological significance of sulphur oxireduction by bacteria. *Philos. Trans. R. Soc. London, Ser. B*, 298: 563–581.

Trudinger, P.A. and Bubela, B., 1967. Microorganisms and natural environment. *Mineral Deposita*, 2: 147–157.

Wake, C.V., Christopher, R.K., Rickard, P.A.D., Anderson, J.E. and Rolph, B.J., 1977. A thermodynamic assessment of possible substrate for sulfate reduction bacteria. *Austr. J. Biol. Sci.,* 30: 155–172.

Ward, D.M., Atlas, R.M., Boehm, P.D. and Calder, J.A., 1980. Microbial biodegradation and chemical evolution of oil from the Amoco Spill. *Ambio*, 9(6): 277–283.

Widdel, F., 1980. *Anaerober Abbau von Fettsäuren und Benzoesäure durch neue isolierten Arten sulfatreduzierender Bakterien*. Dissertation Arbeit, Georg-August Universität zu Goettingen.

Wright, C.C., 1963. The use of CO_2 in waterfloods. *API Prod., Div. Pac. Coast Distr. Meet., May 21-22, Los Angeles, Calif.*

Zajic, J.E., 1969, *Microbial Biochemistry*. Academic Press, New York, N.Y., 345 pp.

Zhang Yi-Gang, 1981. Cool shallow origin of petroleum-microbial genesis and subsequent degradation. *J. Petrol. Geol.*, 3: 427–449.

ZoBell, C.E., 1946. Action of microorganisms on hydrocarbons. *Bacteriol. Rev.*, 10: 1–49.

ZoBell, C.E. and Morita, R.Y., 1957. Barophilic bacteria in some deep sea sediments. *J. Bacteriol.*, 73: 563–568.

Chapter 5

OIL RECOVERY BY BACTERIAL AND POLYMER SOLUTIONS IN THE HELE-SHAW MODEL

J.E. ZAJIC *, W. SEFFENS, A. GURROLA and TAKAYOSHI BAN

INTRODUCTION

Various laboratory models for evaluating oil removal from porous media have been developed. This study deals with using a Hele-Shaw model to test bacterial cultures and products, particularly polymers for oil recovery efficiency. The Hele-Shaw cell, composed of a viscous fluid confined between closely spaced parallel sheets of glass, demonstrates flow instability and decreased swept area if a less viscous fluid is used as the driving fluid. The instability develops into round-ended fingers of less viscous fluid penetrating into the more viscous one. The fluid motion is mathematically analogous to two-dimensional flow in a porous medium (Saffman and Taylor, 1958). The Hele-Shaw model does not purport to be a method of simulating conditions present in a reservoir, but represents a simple method of evaluating and comparing solutions for possible enhanced oil recovery potential using a single-pore system.

Porous media which have been used in evaluating oil separation are: capillary (Fairbrother and Stubbs, 1935), sandstone (Leverett, 1939), Berea sandstone (Taber, 1969; Foster, 1973), teflon, steel and aluminum (Lefebure du Prey, 1973), unconsolidated glass beads, and the Hele-Shaw cell (Saffman and Taylor, 1958). Taber (1969) has given the capillary number and displacement ratio correlating group for each of these systems. In model studies, the conditions for maximum oil recovery are: largest grain size (20–40 mesh); highest porosity (41%); uniform grain size (all 40–48 mesh); intermediate wettability of grains; lowest possible oil viscosity (1 cP); highest temperature used: 125°F (51.7°C); superficial flow rate with additives: less then 1 ft/day (30.5 cm/day); lowest interfacial tension used: 8 mN/m; alkalinity (or salts): 1.0*N* sodium carbonate (Uren and Fahmy, 1927).

---

* Deceased

MATERIALS USED

Oil

A non-detergent motor oil (Industrial Lubricant Company, San Antonio, Texas) was used in experiments.

Viscosity of this oil was determined to be 832 cP and the interfacial tension against distilled water was 27 mN/m. Operational flow through the Hele-Shaw cell was laminar with Reynolds number below 300.

Cultures

Isolates of the genus *Zymomonas*, a Gram-negative bacteria, rod-shaped facultative anaerobe, were used in this study. These cultures produce ethanol and a variety of higher alcohols which have co-surfactant properties. In addition they are anaerobic and produce carbon dioxide as an end product.

Uniform techniques were used in the growth of all strains of *Zymomonas* and all culture broths were passed through the Hele-Shaw model undiluted.

Cultures were grown in a glucose-yeast–salt broth: yeast, 0.5%; K_2HPO_4, 0.1%; $(NH_4)_2SO_4$, 0.1%; $MgCl_2$, 0.05%; $MnCl_2$, 0.0015%; calcium pantoethanoate, 0.001%; biotin, 0.0025%; and glucose, 10%, at pH 7.0. The bacterial flow experiments in the model were completed immediately after a 3-day incubation period at 25°C. The whole broth was used directly.

Polymers

Carboxymethyl cellulose (CMC) polymer and eleven biopolymers were used in this study. Xanthan gum, dextran, heteropolysaccharides, phosphoglucomannans, D-mannans, and phosphomannan-type polymers produced by various microbes were used.

All polymers with the exception of three (xanthan gum, dextran, and CMC) were supplied by Dr. M.E. Slodki (Microbial Biochemistry Research Fermentation Laboratory, U.S. Department of Agriculture) in lyophilized crude form.

Most polymer solutions were evaluated at a 0.2% concentration. The characteristic of the polymers of primary interest to this study was their ability to increase the viscosity of water, thus allowing investigation of mobility control for two-phase flow systems.

APPARATUS

The Hele-Shaw apparatus, illustrated in Fig. 5-1, consisted of (a) a storage vessel with spigot containing the test displacement fluid, rubber, and glass tubing; (b) the Hele-Shaw cell per se; (c) graduated cylinders; and (d) platform for adjusting the

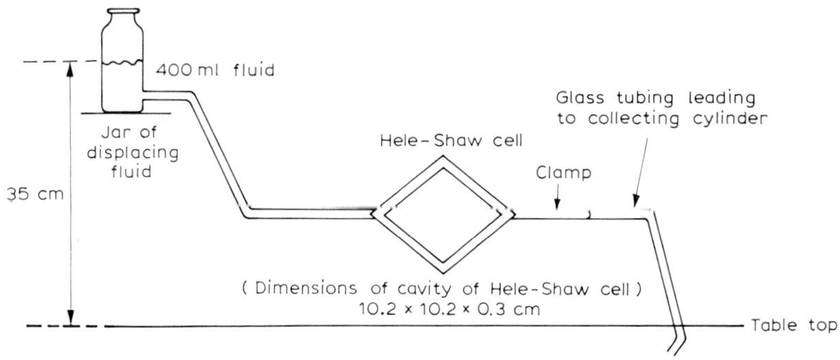

Fig. 5-1. Apparatus and experimental set up of Hele-Shaw model.

height of displacing fluid to achieve a constant head pressure. A second spigoted displacement vessel containing oil was used to fill the Hele-Shaw cell (not shown). A magnetic stirrer was used to mix the displacing fluid to maintain a uniform solution of the fluid. The Hele-Shaw cell per se consists of two closely spaced glass plates (3-mm spacing) with provisions for displacement of fluid from the cell. Spacing of the glass plates was accomplished by placing thin glass strips at edges of the plates and gluing them together with a sealant. Stainless steel tubes were placed at opposite corners of the model for inlet and outlet flow. The internal dimensions were arbitrarily picked to be 10.2 cm × 10.2 cm × 0.3 cm.

EXPERIMENTAL PROCEDURE

The spigoted displacement jar was filled with 400 ml of the fluid to be tested and then raised to a fluid height of 35 cm. This produced a hydrostatic pressure equal to 350 mm of water (density of all solutions used were assumed to be equal to that of water). Care was excercised to remove all air bubbles present in the tubing leading from the spigoted displacement jar. The Hele-Shaw cell was then filled with non-detergent oil by connecting it to the second spigoted displacement jar filled with oil. After filling with oil, the Hele-Shaw cell was connected to the spigoted displacement jar containing test fluid. The oil–water system (the displacing fluid is referred to as "water") was held in place with a clamp at the position shown in Fig. 5-1. 25-ml graduated cylinders were used to collect oil and oil–water flowing through the model.

To initiate a test, the clamp was removed and flow started. As would be expected, the initial displacement of liquid was totally oil. After the point of water breakthrough, the oil/water ratio became extremely small with the ratio decreasing with time. A total of six graduated cylinders were filled (150 ml volume) with displacement fluid. Breakthrough is usually observed in the first 25 ml of fluid volume. The fraction of oil–water remaining in the glass tube was also collected to account for all displaced oil. The rate of fluid flowing through the system was

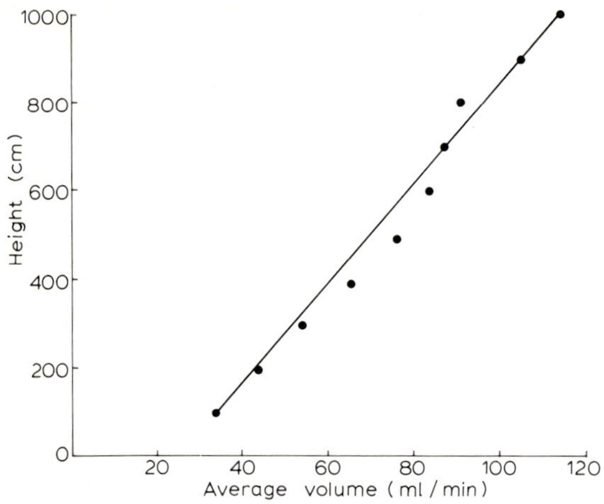

Fig. 5-2. The effect of hydrostatic pressure on the rate of water flow through the Hele-Shaw model.

determined by collecting volumes of fluid and recording times of collection. Test runs were made in duplicate and results for individual runs had good reproducibility.

The Hele-Shaw model has been used to measure displacement efficiency of an aqueous phase pushing an oil phase within a defined pore geometry (Gogarty and Tosch, 1968). Oil displacement ability of bacterial and polymer solutions tested here are discussed in the following sections.

The defined geometry of the Hele-Shaw model allows for characterization of fluid flow through the system. Permeability is defined from Darcy's Law for linear flow of an incompressible fluid in a porous medium:

$$q = kA\Delta p/\mu l \tag{5-1}$$

where k is the specific permeability, q is the flow rate of a fluid, μ is the viscosity, A is the cross-sectional area, Δp is the pressure drop across the medium, and l is the length (see Scheidegger, 1960). The dimension of permeability is length squared, the oil industry using the Darcy (1 Darcy $= 9.87 \times 10^{-9}$ cm^2). Permeability with respect to water in the Hele-Shaw model was found to be 9×10^{-7} cm^2 by finding the slope $\Delta q/\Delta p$ from Fig. 5-2, and using $A = 2.2$ cm^2 and $l = 13.3$ cm as the Hele-Shaw model dimensions (see equation (5-1)). The viscosity of water under test conditions was 10^{-2} g/cm s. The permeability with respect to water calculated for the Hele-Shaw model was comparable to permeability of loose bed sands (Scheidegger, 1960).

Two-phase fluid flow in porous media is characterized as either laminar or turbulent flow. The type of flow is estimated by using the dimensionless Reynolds number:

$$Re = \rho dv/\mu \tag{5-2}$$

where ρ is the fluid density, d is the mean diameter of the porous system, v is the average velocity, and μ is the viscosity of the fluid. The Reynolds number for a flow rate of 50 ml/min and model diameter of 0.4 cm is approximately 250 (equation (5-2)), thus indicating laminar flow through the test system.

The Hele-Shaw model has previously been studied for purposes of investigating dynamics and theory of single- and two-phase flow systems (gas–liquid, liquid–liquid) (Hele-Shaw, 1897, 1898; Taber, 1981). The correlation group functional for evaluation of oil recovery in the Hele-Shaw model according to Saffman and Taylor (1958) is:

$$v\mu/\gamma_i \tag{5-3}$$

where v is the velocity of displacement by wetting phase, μ is the viscosity, and γ_i is the interfacial tension between wetting phase and oil (Saffman and Taylor, 1958). The mobility ratio, M, which is equal to:

$$M = k_1\mu_2/\mu_1 k_2 \tag{5-4}$$

has been used to evaluate oil recovery potential of solutions in the Hele-Shaw model (Mungan, 1971). In the above equation μ and k are viscosity and permeability with respect to fluid 1 (wetting phase) and fluid 2 (non-wetting phase). The Hele-Shaw model was used here to evaluate oil displacement by wetting phase. Interactions affecting the efficiency of oil displacement in reservoir models are readily measured with a Hele-Shaw model. Bacterial culture broths per se and polymers for mobility control have been tested by the writers in the Hele-Shaw model with the following results.

Effect of bacterial cultures

Fourteen strains of *Zymomonas* were examined for MEOR potential in the Hele-Shaw model. Although fourteen isolates were tested, only data on two typical species are presented in Fig. 5-3. All culture broths were tested undiluted. Percentage of recovered oil and efficiency ratings with respect to (a) time, (b) water volume, and (c) time and water volume for these *Zymomonas* species are listed in Table 5-I. These culture broths gave less oil recovery than the control broth (uninoculated media), but were more effective than water controls. Culture broth surface tension, interfacial tension, and viscosity are listed in Tables 5-II and 5-III.

These strains of *Zymomonas* demonstrated little MEOR potential in the Hele-Shaw model. *Zymomonas* cultures were examined for MEOR potential, because they are facultative anaerobes and produce organic acids, carbon dioxide, and alcohols. The organic acids produced could aid in increasing oil reservoir permeability by dissolving carbonate rock (Davis, 1967), the CO_2 can aid in repressurizing the reservoir and in viscosity reduction of oils (La Riviere, 1955), and the alcohols are useful as co-surfactants for use in micro-emulsion flooding (Gogarty and Tosch,

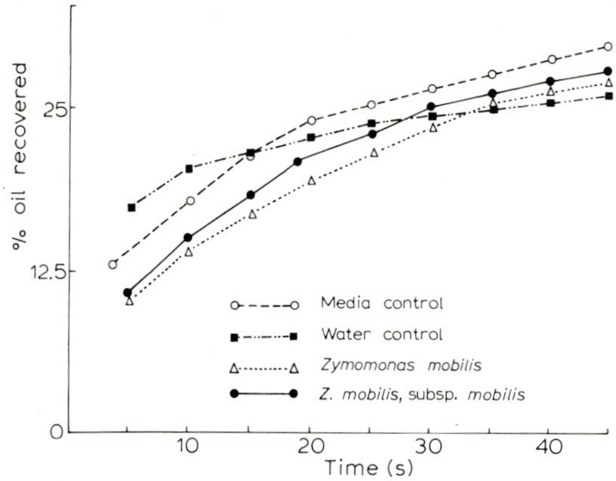

Fig. 5-3. Recovery of oil in the Hele-Shaw model using *Zymomonas mobilis* subspecies *mobilis* (4490) and *Z. mobilis* (8938).

1968). The system used here does not examine the reaction of organic acids with carbonates and the effect of CO_2 in reducing viscosity.

In view of the fact that *Zymomonas* cultures produce large quantities of ethanol from carbohydrate, solutions of ethanol as well as butanol were evaluated for oil displacement with the model (Fig. 5-4). These alcoholic solutions alone had a negative effect in displacing oil from the Hele-Shaw model. A 6% solution of either ethanol or butanol gave γ_i values against oil of 21.3 and 10 mN/m, respectively.

As ethanol and butanol are often oxidized to organic acids of the same chain length, acetic acid and butyric acid were examined (Fig. 5-5). Once again the solutions demonstrated less oil displacement efficiency than water. The γ_i values for

TABLE 5-I
Efficiency ratings, percent oil recovery, surface tension, and viscosity for *Zymomonas* culture used

	ml oil / time (s)	ml oil / "water" vol. (ml)	ml oil / (time) ("water" vol.)
Water	0.23	0.033	0.00073
Z. media control	0.27	0.037	0.00081
Z. 8938	0.23	0.034	0.00069
Z. 4490	0.25	0.038	0.00085
	Recovery (%)	**Surface tension (mN/m)**	**Viscosity (cP)**
Water	26.25	72	11.8
Z. media control	30	64.2	12
Z. 8938	28.25	58.3	13
Z. 4490	28.00	52.8	13

TABLE 5-II

Efficiency ratings and percent recovery of oil for polymer solutions used

Displacing agent	ml oil / time (s)	ml oil / "water" vol. (ml)	ml oil / (time) ("water" vol.)	Recovery (%)
0.8% CMC	0.12	5.36	0.02179	73.7
0.4% CMC	0.21	0.682	0.00541	67
0.2% CMC	0.27	0.374	0.00492	51.5
0.1% CMC	0.35	0.266	0.00502	46.2
0.35% xanthan gum	0.23	0.238	0.00189	72.5
0.17% xanthan gum	0.33	0.156	0.00250	51.7
0.09% xanthan gum	0.35	0.079	0.00253	27.7
0.2% heteropolysaccharide from *Limpomyces starkeyi* (NRRL Y-2543)	0.22	0.23	0.0017	75.5
0.1% heteropolysaccharide from *Limpomyces starkeyi* (NRRL Y-2543)	0.19	0.14	0.0013	40.6
0.2% heteropolysaccharide from *Cryptococcus laurenti* var. *flavescens* (NRRL Y-1401)	0.14	0.30	0.0012	89.8
0.2% O-phosphonoglucomannex from *Pichia pinus* (NRRL Y-2579)	0.19	0.12	0.0014	44.3
0.2% heteropolysaccharide from *Arthrobacter viscosus* (NRRL B-1973)	0.19	0.17	0.0014	57.0
0.2% α-D-mannan from *Torulopsis pinus* (NRRL Y-2023)	0.20	0.12	0.0014	44.3
0.2% α-D-mannan from *Hansenula holsti* (NRRL Y-2448)	0.19	0.12	0.0013	44.8
0.2% α-D-mannan from *Hansenula capsulata* (NRRL Y 1842)	0.19	0.13	0.0013	45.3
0.2% O-phosphonomannan from *Hansenula holsti* (NRRL Y-2448)	0.20	0.16	0.0015	52.8
0.2% α-D-glucomannan from *Pichia pinus* (NRRL Y-2579)	0.25	0.12	0.0017	44.8
0.2% dextran from *Leuconostoc mesenteroids*	0.18	0.12	0.0012	43.3
0.2% carboxymethyl cellulose	0.15	0.22	0.0012	71.5
Water	0.22	0.11	0.0015	39

6% solutions of acetic and butyric acid against oil were 19.8 and 12 mN/m, respectively. *Zymomonas* cultures also produced CO_2. Figure 5-6 shows results obtained using dissolved CO_2 in solution. Percent concentrations of dissolved CO_2 in solutions were determined via a phenolpthalein–Na_2CO_3 titration system (Feigl, 1958). Oil displacement efficiency is seen to be concentration dependent, with solutions containing low concentrations of CO_2 demonstrating poor enhanced oil recovery and solutions of higher concentration demonstrating slight improvement in

TABLE 5-III
Viscosity, surface tension, and interfacial tension for polymer solutions used

	γ_s (mN/m)	γ_i (mN/m)	Viscosity (cP)	Temperature (°C)
A. Non-detergent oil	34.4	–	832	25
B. Polymers:				
0.2% heteropolysaccharide from *Limpomyces starkeyi* (NRRL Y-2543)	67.7	23	70.8	23.4
0.1% heteropolysaccharide from *Limpomyces starkeyi* (NRRL Y-2543)	64.8	22.3	24	23.3
0.2% heteropolysaccharide from *Cryptococcus laurenti* (NRRL Y-1401)	75.9	27.3	114.8	24.5
0.2% O-phosphonoglucomannan from *Pichia pinus* (NRRL Y-2679)	62.9	20.2	19.6	24.4
0.2% heteropolysaccharide from *Arthrobacter viscosus* (NRRL B-1973)	64.2	22	34.4	24.2
0.2% α-D-mannan from *Torulopsis pinus* (NRRL Y-2023)	63	17.6	18.4	24.4
0.2% dextran	74.5	25.6	17.2	24.5
0.2% α-D-mannan from *Hansenula holsti* (NRRL Y-2448)	67.6	23.7	14.4	24.5
0.2% α-D-mannan from *Hansenula capsulata* (NRRL Y-1842)	69.6	21.2	14.4	24.6
0.2% O-phosphonomannan from *Hansenula holsti* (NRRL Y-2448)	68.2	22.2	30.4	24.5
0.2% α-D-glucomannan from *Pichia pinus* (NRRL Y-2579)	62	23.1	19.4	24.4
0.2% CMC	75.1	17	81.6	24.1
0.4% CMC	75.7	12.1	140	25.7
0.17% xanthan gum	47.2	11.0	40	25.0
0.35% xanthan gum	6.3	–	82	25.0

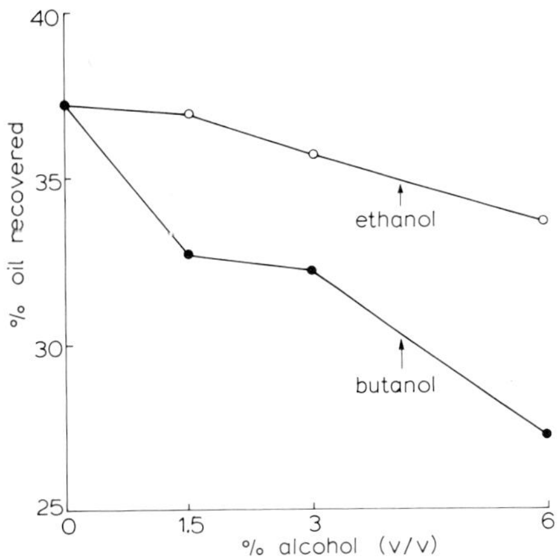

Fig. 5-4. Recovery of oil in the Hele-Shaw model by alcohol solutions.

oil recovery as compared with water (0.08–0.10% dissolved CO_2 solutions). These effects could be based in part on the correlation group of equation (5-3), indicating that oil recovery is a function of interfacial tension and flow velocity. Saffman and Taylor (1958) found that interfacial tension effects became important for very slow flow rates by influencing the profile of the driving plug.

Effect of polymer solutions

Polymers have a significant impact upon increasing the viscosity of water. Mobility control, which is greatly influenced by viscosity, is an important factor

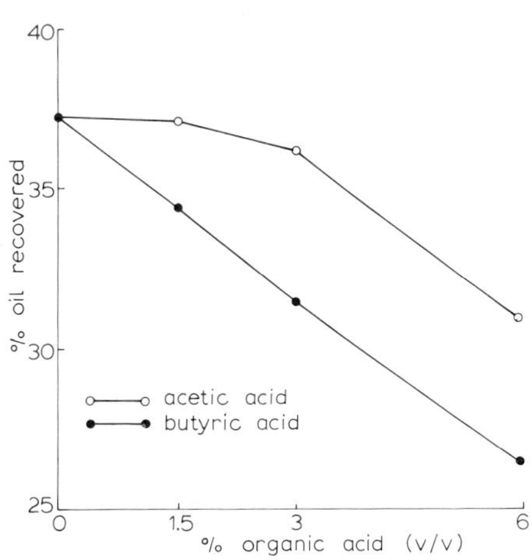

Fig. 5-5. Recovery of oil in the Hele-Shaw model by organic acid solutions.

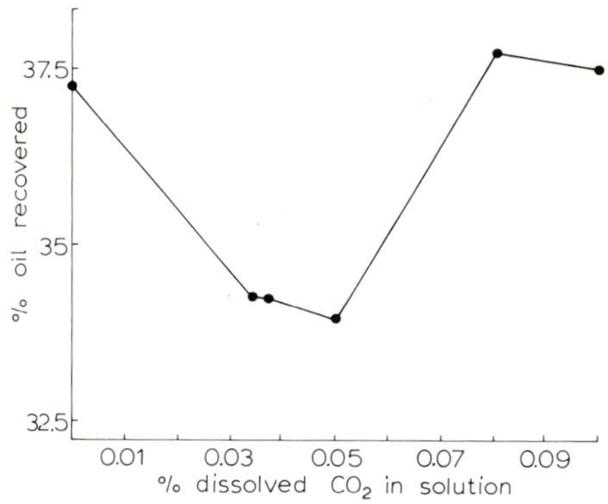

Fig. 5-6. Recovery of oil in the Hele-Shaw model by CO_2 solutions.

when considering flooding as a means of enhancing oil recovery (Mungan, 1971). Analyzing oil displacement using mobility ratios (equation (5-4)), radial encroachment at a well has been investigated (Muskat, 1934). The position of the interface is denoted by r_o and assumed to be circular at all times. Boundaries of the well and encroaching fluid are denoted as circles of radius r_w and r_e ($r_w < r_o < r_e$), respectively. Assuming $r_e/r_w = 2000$, one finds that mobility ratios less than or equal to 1 are most efficient at displacing oil. This demonstrates that higher mobility ratio systems give a slower displacement, especially for M values above 1. The M values below 1 showed increased displacement efficiency. Displacement efficiencies, which are volumes of displaced fluid relative to the pore volume, can be examined in a more complicated model as in the Hele-Shaw model (Hele-Shaw, 1897, 1898). Permeability values with respect to water and oil were calculated for the Hele-Shaw model. Assuming that all aqueous phase solutions have a permeability equal to that of water, mobility ratios were calculated and are presented in Tables 5-IV and 5-V for various solutions tested in the Hele-Shaw model. As the more viscous fluids

TABLE 5-IV

Mobility ratio determination for various solutions tested in Hele-Shaw model [a]

Aqueous phase	Viscosity (cP)	Calculated mobility ratio	Recovery (%)
Water	10	5.35	26.25
Zymomonas (10988)	12	4.46	26.25
0.2% heteropolysaccharide from *Arthrobacter viscosus*	34.4	1.55	57.0
0.2% CMC	81.6	0.66	51.5
0.2% heteropolysaccharide from *Cryptococcus laurenti*	114.8	0.47	89.8

[a] $M = k_1\mu_2/k_2\mu_2$; $k_w/k_o = 0.064286$; $\mu_o = 832$ cP.

TABLE 5-V
Correlation group values for various solutions tested in Hele-Shaw model

Aqueous phase	Calculated correlation group, equation (5-3)	Oil recovery (%)
Water	0.36	26.25
Zymomonas medium control	0.31	30
Zymomonas 4490	0.36	28.25
Zymomonas 8936	0.47	28
0.17% xanthan gum	2.02	51.75
0.35% xanthan gum	4.71	72.5
0.4% CMC	8.49	67

demonstrated lower mobility ratios, they would be expected to show better oil recovery ability. Observation of percentage oil recovered supports this conclusion in line with the results of others. The shape of the meniscus formed at the interface between water and oil is a function of the correlation group (Saffman and Taylor, 1958) in question (Taber, 1969). The shape of the meniscus is directly proportional to the amount of oil displaced (the larger the meniscus, the more oil displaced); therefore, the correlation group can be used to evaluate displacement efficiency.

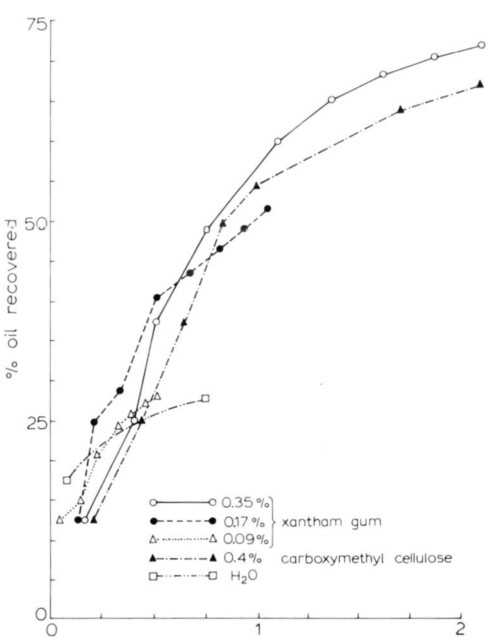

Fig. 5-7. Recovery of oil in the Hele-Shaw model using xanthan gum solutions and carboxymethyl cellulose solutions.

Inasmuch as conditions, which favor oil displacement, are high flow rates, high viscosity, and low interfacial tension, the correlation group should have a large value.

Flow rates are a function of the viscosity of the displacement fluid and were determined for each solution investigated.

Using equation (5-3), correlation group values for various solutions tested are presented in Table 5-V. Observation shows that high correlation group values are paired with the fluids that demonstrated the greatest percent recovery of oil. Viscosity seemed to be the predominant factor affecting the correlation group.

Eleven biopolymers and carboxymethyl cellulose were examined using the Hele-Shaw model for enhanced oil recovery potential. The purpose was to test influence of viscosity (high viscosity) upon oil recovery. Table 5-III contains data on viscosities, γ_s and γ_i (interfacial tensions) of the different polymers.

Fig. 5-7 examines results of testing polymer solutions of xanthan gum and CMC at various concentrations (i.e., different viscosities). In both instances, oil recovery increased with increasing concentration (i.e., higher viscosities), but with very high viscosity systems fluid velocity and time became controlling factors. Fig. 5-7 also compares oil recovery between CMC and a biopolymer (xanthan gum). A 0.35% concentration of xanthan gum gave better oil displacement results than 0.4% CMC solution.

Results obtained in examining the remaining microbial polysaccharides are shown in Fig. 5-8. The heteropolysaccharide from *Cryptococcus laurenti* var.

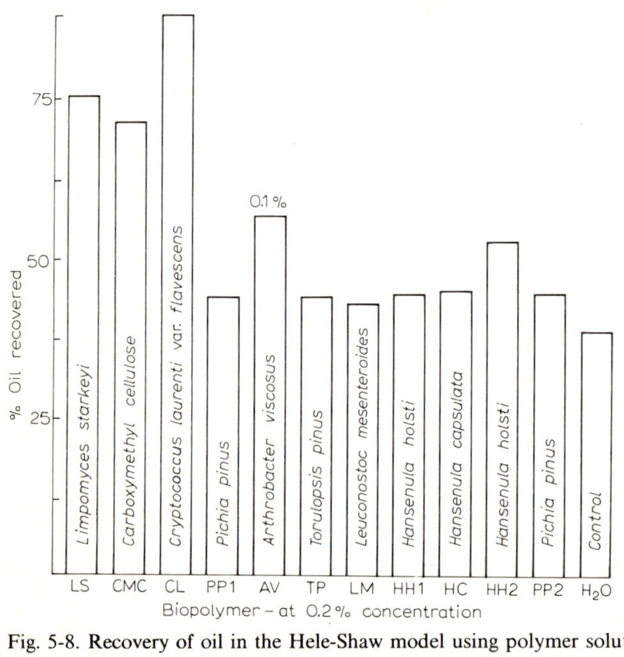

Fig. 5-8. Recovery of oil in the Hele-Shaw model using polymer solutions at a 0.2% concentration.

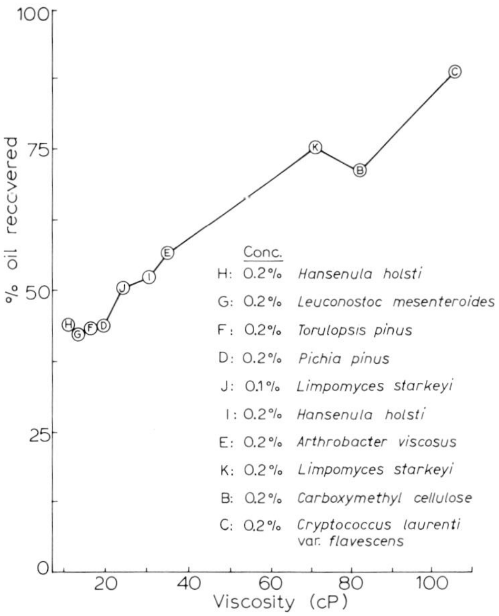

Fig. 5-9. Recovery of oil in the Hele-Shaw model using fluids of varying viscosities.

flavescens gave best oil displacement under the conditions used. Of the remaining biopolymers examined, the heteropolysaccharides demonstrated the best oil recovery ability, with two giving better oil displacement than water. The actual percentage of oil recovered and efficiency ratings for the biopolymers tested are listed in Table 5-II.

All biopolymers were tested at 0.2% concentration (weight/volume). At this concentration, viscosities were also evaluated and reported. The relationship between viscosity of displacing fluid and oil recovery can be seen in Fig. 5-9 (in this comparison, polymer type is de-emphasized). The most viscous fluids resulted in the greatest oil recovery, demonstrating the importance of the viscosity in the correlating group. The polysaccharides giving the greatest viscosities per unit weight are the best in oil displacement from the model. The viscosity, μ, and interfacial tension, γ_i, values for these polysaccharides evaluated are presented in Table 5-III. None of the biopolysaccharides possessed attractive surface tension lowering properties.

CONCLUSIONS

All strains of *Zymomonas* gave about the same level of oil displacement in the Hele-Shaw model. Their performance was less than that observed for the growth medium alone, but better than for waterflood controls. The interfacial tension values produced by 6% ethanol, butanol, acetic acid and butyric acid were not adequate to influence the interfacial tension value of the correlation group to give good oil displacement.

Viscosity of a displacing fluid was determined to be an important factor in increasing oil recovery in the Hele-Shaw model, particularly because of its contribution to mobility control. High-viscosity heteropolysaccharide solutions from organisms, such as *Cryptococcus laurenti* and *Lympomyces starkeyi*, performed most efficiently in the Hele-Shaw model. The higher the viscosity of a solution, the greater was its ability to displace oil.

The correlation group indicates that relatively high velocities and viscosities are required to displace oil. The interfacial tension values must be as low as possible.

REFERENCES

Davis, J., 1967. *Application of Microbiological Methods for Secondary Oil Recovery from Carpathian Crude Oil Reservoirs*. Elsevier, Amsterdam.

Fairbrother, F. and Stubbs, A.E., 1935. Electroendosmosis, VI. "Bubble-type" method of measurement. *J. Chem. Soc.*, pp. 527–529.

Feigl, F., 1958. *Spot Tests in Organic Analysis*. Elsevier, Amsterdam, 337 pp.

Foster, W.R., 1973. A low-tension waterflooding process. *J. Pet. Technol.*, 25(2): 205–210.

Gogarty, W. and Tosch, W., 1968. Miscible-type waterflooding: oil recovery with micellar solutions. *J. Pet. Technol.*, 20(12): 1407–1414.

Hele-Shaw, H., 1897. *Trans. Inst. Naval. Arch.*, 39: 145.

Hele-Shaw, H., 1898. Investigation of nature of surface resistance of water and of stream-line motion under certain experimental conditions. *Trans. Inst. Naval. Arch.*, 40: 21.

Karashiewiez, J., 1974. *Pr. Inst. Naft., Cracow*.

La Riviere, J., 1955. Production of surface active compounds by microorganisms and its possible significance in oil recovery. *Antonie van Leeuwenhoek; J. Microbiol. Serol.*, 21: 9–27.

Lefebvre du Prey, E.J., 1973. Factors affecting liquid–liquid relative permeabilities of a consolidated porous medium. *Soc. Pet. Eng. J.*, 13: 39–47.

Leverett, M.C., 1939. Flow of oil–water mixtures through unconsolidated sands. *Trans. AIME*, 132: 149.

Mungan, N., 1971. Improved waterflooding through mobility control. *Can. J. Chem. Eng.*, 49: 32–37.

Muskat, M., 1934. Flow of compressible fluids through porous media and heat conduction. *Physics*, 5: 71–94.

Saffman, P.G. and Taylor, G., 1958. Penetration of a fluid into a porous medium or Hele-Shaw cell containing a more viscous liquid. *Proc., R. Soc. London, Ser. A*, 245: 312.

Scheidegger, A.E., 1960. *Physics of Flow Through Porous Media*. University of Toronto Press, Toronto, Ont., pp. 73–98.

Taber, J.J., 1969. Dynamic and static forces required to remove a discontinuous oil phase from porous media containing both oil and water, *Soc. Pet. Eng. J.*, 9: 3–12.

Taber, J.J., 1981. Research on enhanced oil recovery: past, present and future. In: D. Shah (Editor), *Surface Phenomena in Enhanced Oil Recovery*. Plenum Press, New York, N.Y., pp. 13–52.

Uren, L.D. and Fahmy, E.H., 1927. Increasing oil recovery by flooding, 1. Theoretical conditions. *Trans. AIME*, 77: 318.

NOTE ADDED IN PROOF

The γ_i values against *n*-hexadecane for 6% aqueous solutions of ethyl alcohol, *n*-butyl alcohol, acetic acid, and *n*-butyric acid were determined to be 27.6, 4.4, 26.7, and 10.5 dynes/cm, respectively.

Chapter 6

OIL DISPLACEMENT BY ANAEROBIC AND FACULTATIVELY ANAEROBIC BACTERIA

E.A. GRULA *, H.H. RUSSELL, D. BRYANT and M. KENAGA

INTRODUCTION

The use of bacteria in oil recovery (MEOR) is not a new idea. What may be considered new relates to attempts to select and isolate organisms such as certain species of *Clostridium* from nature that have attributes considered beneficial or essential to the process (Grula et al., 1983).

ZoBell (1946) was the first to seriously advocate use of bacteria in oil recovery. He proposed and, indeed, was issued a patent in the area. He proposed that bacteria could release oil if they possessed some or all of the following characteristics: (1) ability to decompose oil with consequent production of organic acids and gases **; (2) dissolution of carbonate rock by the produced acids; (3) a decrease in specific gravity and viscosity of oil because of carbon dioxide production and its solubility in the oil; (4) increased reservoir pressure due to the microbial generation of hydrogen, methane and carbon dioxide; and (5) decrease in surface tension due to formation of surfactants. In addition, growth of bacteria upon the rock surfaces would bring about physical dislodgement of attached oil. According to ZoBell (1946, 1947a) an organism ideally suited for use in MEOR was *Desulfovibrio hydrocarbonoclasticus* ***. Apparently, no successful field experiments have ever been conducted using this or other species of *Desulfovibrio* (Davis, 1967).

Experiments conducted in the United States considered successful in release of oil by bacteria have been reported by Coty (1976) and Johnson (1979). The experiment of Coty and his colleagues, Updegraff and Yarbrough, was an extensive line-drive type experiment conducted in Union County, Arkansas, using a laboratory strain of *Clostridium acetobutylicum*. Numerous periodic feedings were done using molasses plus certain salts.

---

* Deceased September 1, 1983.

** This process generally requires organisms possessing oxygenases and the presence of molecular oxygen; the normal anaerobic conditions in an oil reservoir would preclude such activity. Acids and gases could be produced anaerobically from fermentation of an added substrate such as molasses, but not from oil.

*** This is not a valid species name. There is no evidence that any species of *Desulfovibrio* can degrade hydrocarbons (Postgate, 1979).

Three months after injecting molasses–salts along with bacteria, a statistically significant increase in oil production began, and continued for seven months, after which the experiment was terminated. Based on the assumed normal decline curve, the average oil production rate for the period November 1954 to May 1955 should have been approximately 0.6 bbl/day, whereas the actual average production was 2.1 bbl/day, 3.5 times the "normal". Considering the very low level of residual oil ($<10\%$) in the formation, the experiment has to be considered a success. It appears that the low price of crude oil in the 1950–1960 period held up further development of the technology. *Clostridium acetobutylicum* with molasses–salt feeding has also been used by Johnson in numerous huff-and-puff-type field experiments. It is claimed by Johnson (1979) that a significant increase in crude oil production (in excess of 350%) was achieved in a fairly large number of (about 150) treated wells (see also Hitzman, 1983). Excellent results have been claimed in many field trials and an excellent summary of these tests has been made by Hitzman (1983). Practically all of these field tests were done in Eastern Europe (Poland, Czechoslovakia, Hungary, and Romania) and the U.S.S.R.

A patent for use of spore-forming microorganisms in MEOR was granted to Hitzman in 1962. Both aerobic and anaerobic organisms were included; the only requirement was the presence of a spore stage that can readily be changed to the metabolizing stage by the addition of proper nutrients. In addition to recommending use of spores rather than vegetative cells for inoculation, it was pointed out that these microorganisms produce organic acids, enzymes, gases and wetting agents from carbohydrates. Increased mobility of unsaturated compounds in the formation was postulated to occur because of possible splitting of double bonds in such molecules by the metabolic activity of the microorganisms.

Because some good results had been obtained in the Mobil field test (Yarbrough and Coty, 1983), in which laboratory cultures of *C. acetobutylicum* were used, the writers set out to obtain isolates of *Clostridium* from natural sources capable of producing, in pure or mixed culture, large amounts of desired metabolic end-products (gases, preferably CO_2, short-chain organic acids, solvents such as acetone and low-molecular-weight alcohols, and emulsifier compounds) in a molasses–salts medium without biodegradation of in-situ crude oil. In addition to producing the desired metabolic end-products, clostridia are excellently suited for use downwell because they are anaerobic and, also, produce a hardy endospore to help ensure survival under adverse conditions. An additional reason for utilizing spores for injection is that, unlike vegetative bacterial cells, they are free of an electric charge which might cause them to adhere to rock surfaces, thus hindering their passage through the reservoir. The writers believed that the reservoir environmental conditions under which isolates should work included a salt concentration (as NaCl) of no more than 7.5%, a temperature range of 20–50°C and a depth of no more than 3000 ft. Below this depth, the temperature could become a limiting factor.

Isolation of clostridial species was accomplished using soil, water and other types

TABLE 6-I
Effects of 5% NaCl on various metabolic properties of certain strains of *Clostridium*

Property affected	Percent reduction in presence of 5% NaCl			
	strain 69E	51B	69E+51B	HR-32
Acetic acid production at 48 hr	77.8	85.4	79.5	–
CO_2 production at 48 hr	50	40	44.0	–
Uptake of sucrose 5 min exposure	–	–	–	75.7

of source materials from many diverse places. Significant numbers of good isolates have been obtained that are capable of producing one or more of the desired end-products singly or in combination with other isolates while growing in various molasses media. There is no one source for isolation that can be considered better than any other.

Early in the program, the writers became aware that salt concentrations of 5% or greater would be deleterious to growth of clostridia and to production of the desired metabolic end products. It is an established fact of microbiology, and well known even to ancient civilizations, that presence of salt (as NaCl) at levels of 5% or greater functions as an excellent preservative to inhibit the spoilage of various foods.

Many clostridia can be isolated that yield large amounts of the desired metabolic end-products in the presence of 2.5% or less of NaCl. Unfortunately, when the level in test media is raised to 5.0–7.5%, the number of isolates that can perform favorably (growth and production of desired metabolic end-products) is significantly diminished. Data showing the effects of 5% NaCl on selected clostridia are given in Table 6-I. Depending on the strain, it may be that only the *rates* of growth and production of metabolic end-products are affected. Thus, with some isolates, it should be possible to achieve the same ultimate level of production simply by increasing time of incubation.

Other effects of NaCl concentration in the range of 5.0–7.5% have been observed, one being growth of individual cells into a filamentous (non-dividing) form. Normally, vegetative cells of clostridia are 3–5 μm in length; however, some species will grow in lengths of 50–100 μm or greater in 5.0–7.5% NaCl solutions. Secondly, some isolates tend to agglutinate and grow in large masses. Thirdly, some produce copious amounts of a sticky slimy material, again leading to growth of large cohering masses of cells. Regardless of which of the above occurs at high salt concentrations, plugging of the well could be the ultimate, predictable outcome.

Because not all bacteria respond to high salt concentrations in the same way, a general rule cannot be set forth. Additionally, complexity of the growth medium has much to do with response of clostridia to high salt concentration. Some data to

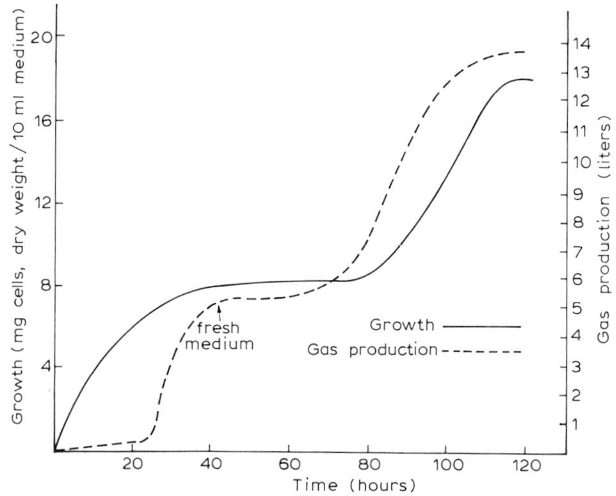

Fig. 6-1. Gas production by strain HR-3E.

demonstrate this type of behavior are included in Table 6-I. Based on observations of the writers it is apparent that all isolates must be individually screened under several conditions of growth before attempting to predict their behavior in an oil well.

It is generally accepted that acid- and solvent-producing clostridia first form acids from carbohydrates. The resulting drop in pH is believed to then activate latent solvent-producing enzymes responsible for reductive synthesis of solvents from their analogous acid form. Synthesis and excretion of acids continues in strain HR-3; however, as also shown, pH of the spent medium remains at another constant value. This is probably because no further ionization of the excreted acids takes place (pH is below the pK of the acids produced). Representative data, included in Figs. 6-1 to 6-5 show that initiation of solvent synthesis by HR-3 is not directly dependent on the attainment of a low pH or the lowest pH the culture will reach. Such data are similar to those presented by George and Chen (1983).

When the writers initiated these studies, no data were available regarding production of emulsifiers by clostridia or other anaerobes from non-hydrocarbon precursors. Investigators have commonly utilized aerobic bacteria plus hydrocarbons as necessary precursors (Cooper and Zajic, 1980). Cooper et al. (1980) reported that *C. pasteurianum* will produce an emulsifier under anaerobic conditions in the absence of any hydrocarbon precursors (molasses–salts–yeast extract medium was used).

Screening of the more promising isolates has shown that many clostridia will produce emulsifier compounds under anaerobic conditions in molasses–salts media (Grula et al., 1983). The lowest surface tension reading thus far obtained using

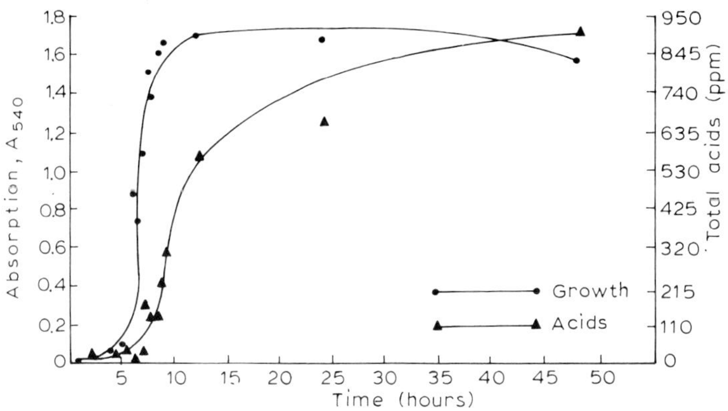

Fig. 6-2. Strain HR-3: growth and acid production vs. time. Cell mass, mg/ml, is proportional to absorbance.

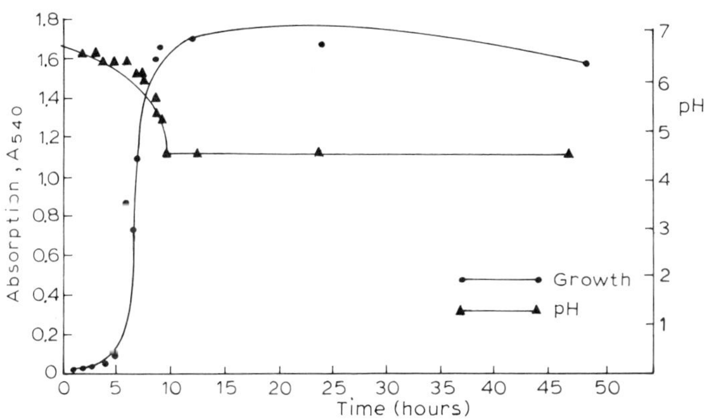

Fig. 6-3. Strain HR-3: growth and pH vs. time.

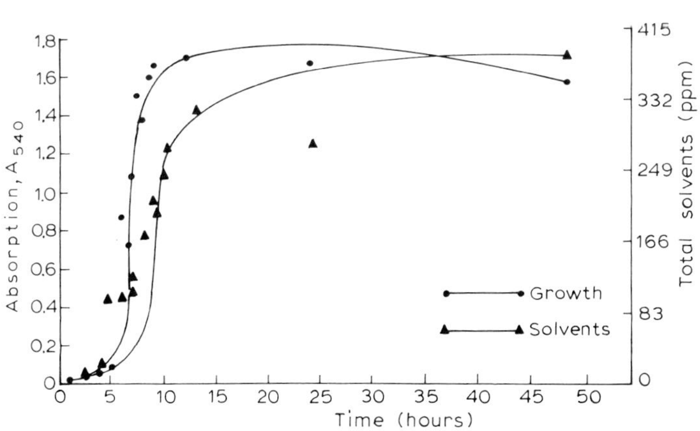

Fig. 6-4. Strain HR-3: growth and total solvent production vs. time.

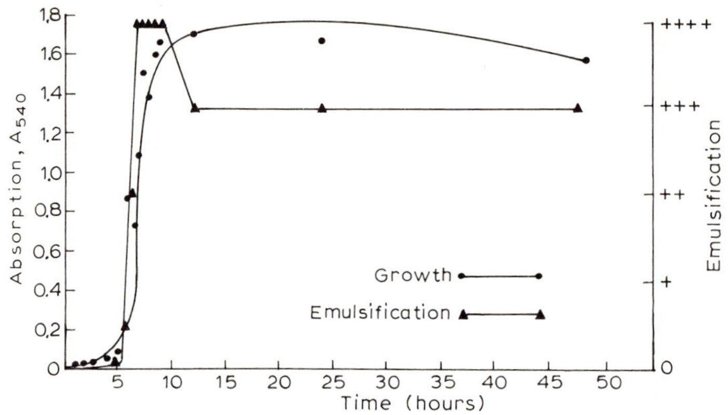

Fig. 6-5. Strain HR-3: growth and emulsifier production vs. time.

unconcentrated culture medium was 45–55 mN/m prior to concentration by foaming (initial surface tension of the uninoculated medium is 60–68 mN/m). Cooper et al. (1980) reported that the surface tension of spent medium after growth of *C. pasteurianum* was 50–60 mN/m prior to foaming.

As shown in Fig. 6-5, production of exocellular emulsifier can be fairly rapid and appears to coincide with the onset of exponential growth. Such synthesis differs from that observed using an aerobic organism such as *Acinetobacter calcoaceticus* (Pines et al., 1983) wherein exocellular production of emulsan, a polyanionic D-galactosamine-containing emulsifier, appears to be related to sloughing of capsular material during the stationary phase of growth. Although the emulsifier material produced by *Clostridium* HR-3 has not yet been chemically characterized, it is heat-stable (withstands autoclaving at 121°C for 10 minutes), probably non-ionic (not precipitated by Ca^{2+}) and generally lacks toxicity toward other bacterial species.

In addition to synthesis of exocellular emulsifiers, it has been possible to show that cells of *Clostridium* HR-3 possess good emulsifier activity even after 3 washes in saline solutions. Emulsions are formed that can remain intact for 7–10 days. Such behavior plus the absence of any microscopically visible capsular or slime material leads to the conclusion that the clostridial cell wall surface is responsible for cellular emulsifying ability. This property of the cell surface will aid in oil recovery because, in addition to ordinary physical displacement of crude oil from solid surfaces by growth, bacterial cells can be directly involved in emulsifying activity. Ability of bacterial surfaces to act as emulsifiers is not restricted to clostridia. Although only minimal testing has been done, the writers noted that laboratory stock cultures of organisms such as *Staphylococcus aureus* produce little or no exocellular emulsifiers; however, washed cells of this organism allow formation of emulsions that are stable for weeks.

Lloydminster Oilfield in Canada

The ultimate recovery by primary and secondary methods from the Lloydminster Oilfield in Canada (Alberta/Saskatchewan border) is estimated to be no higher than 8%. The oil in place is about 2.5 billion cubic meters at a depth of 600 m with an API gravity range of 13–17°. Hence, this is a significant target for MEOR. The principal impediment to production of the oil is water channeling due to the high viscosity of the oil (4000–9000 cP at 25°C), high permeability (4 D), and poorly consolidated sand. Jack et al. (1983) suggested that there are two approaches to MEOR that can have a significant impact on the oil recovery efficiency: (1) repressurization with gases produced by microbes in situ, and (2) release of oil in portions of the reservoir by anaerobic fermentation of molasses which may cause in situ production of acids, surfactants, and solvents. Production of these metabolic products depends on the type of bacteria, the type of nutrients, and the immediate environmental conditions. Preliminary laboratory experiments conducted by Jack et al. (1983) indicated that the major mechanism for oil release from the sand is the nascent formation of gas in situ which brings about a marked decrease of oil viscosity accompanied by swelling as the nascent gas dissolves in the oil. The effect is very pronounced in the laboratory, but may not be as efficient under field conditions because of low gas transfer rates and the absence of convective mixing.

According to Jack et al. (1983), emulsification of the viscous crude oil in situ is not feasible at this stage of development because transport of bacteria within the formation and mixing of the emulsifier (biopolymers?) with the oil-in-place are not technically feasible. However, emulsification of a produced heavy oil as it is pumped from the wells to storage tanks or in the tanks by active microbes, would facilitate later transfer from the tanks, and in pipelines.

Field applications in China

Zhang and Qin (1983) stated that because the oilfields that have thus far been discovered in China are very heterogeneous and contain viscous oils, the approach to MEOR in China has been the development of microbial cultures that can utilize crude oil hydrocarbons to make bioproducts which may aid in recovery of oil as waterflood additives, preparation of drilling fluids, and other oilfield operations. They have found that aerobic fermentation of crude oil yields a biosurfactant which readily forms a microemulsion with crude oil and water. This microemulsion has a greatly reduced viscosity compared to the crude oil. When the products of fermentation are mixed with crude oil having a viscosity of 2500 cSt, in a 1 : 1 ratio, the viscosity of the resulting mixture is between 12 and 46 cSt. Zhang and Qin (1983) postulated that biosurfactants of this type will aid in recovery of residual oil when added to injected water; this will be the emphasis in the future in China.

Zhang and Qin (1983) briefly described a second microbial product which they called a thickening agent because its intended use is to plug high-permeability zones by injection with water. It is a viscous polysaccharide obtained from a Gram-positive, non-spore forming rod with a tendency toward V-shaped association. They have tentatively named the microbe *Corynebacterium gummiferm*. This organism produces copious yields of polymer from mixed paraffins as its sole carbon source, but shows a considerably reduced yield when grown with a crude oil substrate.

The thrust of future work in MEOR in China will be to develop technology for the production of the biosurfactant and biopolymer in above-ground fermentations in the oilfield where they will be used as additives to injection water. Some workers in the field regard this approach simply as a method of chemical flooding (D. Hitzman, personal communication). True MEOR involves in-situ activities of living organisms in the oil reservoir, where their metabolism results in conditions favoring the recovery of oil.

Field applications in Romania

MEOR research in Romania was carried from the laboratory stage directly to field trials with work started in 1972 by Lazar and co-workers (Lazar, 1983). The work began with a concentrated effort to isolate bacteria that were adapted to reservoir conditions. Initially, bacteria were isolated from produced water, but the search was widened to include well bottomhole mud, soils around oil wells, food processing wastes, and others. Oil displacement experiments showed conclusively that the microbial cultures obtained from produced waters and sugar processing plants were more effective. The mixed cultures were adapted to reservoir conditions * and grown to concentrations of 10^7–10^9 bacteria/ml. Genera identified in these mixed cultures were *Pseudomonas, Escherichia, Arthrobacter, Mycobacterium, Micrococcus, Peptococcus, Bacillus*, and *Clostridium*. These mixed cultures were more efficient in releasing oil than the pure strains or bacteria obtained from other sources.

Seven oilfields in Romania were inoculated, but only two of the oilfields responded with increases of production, ranging from 16 to 200 %. The well with the 200% increase had produced 6.7 m^3/month over a period of several years before inoculation and then average production rose to 20.0 m^3/month after inoculation of the field. The increased oil production continued for more than one year and is

* As this work was done before the days of genetic engineering, it is not known how the adaptation took place; a possibility is mutation and selection using either naturally occurring or induced mutations.

currently rising. Lazar attributes the failure of the other five reservoirs to respond to microbial treatment to heterogeneity, low permeability, and high salt concentration.

Monitoring of the microbial population in the produced waters after inoculation showed that about six months after inoculation the microbial count began to increase from 10^4 to 10^6–10^9/ml while a nutrient solution of 2% molasses was being injected constantly. One year after the nutrient injection stopped, the microbial counts in the produced waters dropped to 10^3–10^4/ml. The results indicate that the inoculated microbial culture multiplied in the reservoir and migrated from the injectors to the producing wells.

Field test in Arkansas

Yarbrough and Coty (1983) described a successful field test using *Clostridium acetobutylicum* in a two-spot pattern with 120 m between wells. The formation is a loosely consolidated sand of high permeability containing up to 16% carbonates at a depth of 700 m. Core tests indicated a maximum residual oil saturation of 8.5% and the test well production averaged 3 m^3/month prior to the inoculation.

Injection of a 2% molasses solution and bacteria commenced in July 1954 and was continued until November at an average rate of 25 m^3/day; a total of 800 liters of bacterial culture was injected in this period. In October, 3 months after the initial injection, significant changes occurred at the production well. The oil recovery increased from 3 to 10 m^3/month and continued through the duration of the test (May 1955). Production of acids, carbon dioxide, and methane also began at the production well in October 1954. The cumulative production of acids and carbon dioxide over the test period was 35,000 kg and 5700 m^3 (11,000 kg), respectively. No quantitative measurement of the methane production was made. Hydrogen was not detected, and the authors believe that it may have been used by other bacteria in the reservoir and was involved in the production of much of the methane. It is generally agreed that clostridia do not produce methane; however, some species do produce hydrogen in the laboratory. Since hydrogen was not detected in the experiments of Yarbrough and Coty (1983), and to knowledge of the writers has not been reported in the analysis of gases produced in any downwell MEOR test, it is entirely possible that in the reservoir environment it is used as rapidly as it is formed by methanogenic bacteria to produce methane from carbon dioxide. Ivanov and Belyaev (1983) demonstrated the presence of methanogens using a Most Probable Number type of count, and methanogenesis, in highly freshened oilfield brine.

SUMMARY

The findings presented in this chapter can be summarized as follows:

(1) Salt-tolerant bacteria of the genus *Clostridium* have been isolated and shown to produce relatively large amounts of carbon dioxide and low-molecular-weight

organic solvents in 5% salt solutions. Spores of these bacteria exhibit very little adsorption on sandstone rock samples; thus, the spores can penetrate deep within a petroleum reservoir to germinate where most needed.

(2) A mixed culture has been developed which produces a biosurfactant when grown aerobically with crude oil that causes a decrease of the oil phase viscosity of 95% and produces a non-wetting (on glass or steel) emulsion. The bacteria were shown to utilize the paraffins of the oil to make the extracellular biosurfactant.

(3) Laboratory experiments indicate that some bacteria can be injected into high-permeability zones to increase the sweep efficiency of a waterflood.

(4) Living cells have been found to be strongly adsorbed on sandstone which restricts the injection of non-spore-forming bacteria unless carrier solutions are developed that inhibit cell adsorption and thus allow deep migration of the cells into the oil reservoir.

(5) Mixed bacteria cultures (Gram-negative, facultative rods) have been isolated that produce copious amounts of gas. These have been designated for possible repressurization of oilfields in Canada. A nutrient solution that aids in the injection of the bacteria by inhibiting adsorption also was developed. A second growth medium activates the bacteria to produce a biopolymer in situ.

(6) An extracellular polysaccharide prepared from aerobic fermentation of crude oil has been developed in China and is proposed as a waterflood additive to increase sweep efficiency and oil recovery.

(7) Two of seven field tests in Romania with mixed cultures of bacteria that were acclimated to reservoir conditions resulted in a two-fold increase of production. Oil production continues to rise in the two fields that responded positively.

(8) A field tests in the United States with *Clostridium acetobutylicum* resulted in a two and one-half fold increase in production from a two-spot injection test and production of 34,000 kg of acids and 5700 m^3 of carbon dioxide during the 9-month test period.

REFERENCES

Cooper, D.G. and Zajic, J.E., 1980. Surface-active compounds from microorganisms. *Adv. Appl. Microbiol.*, 26: 229–253.

Cooper, D.G., Zajic, J.E., Gerson, D.F. and Manninen, K.I., 1980. Isolation and identification of biosurfactants produced during anaerobic growth of *Clostridium pasteurianum*. *J. Ferment. Technol.*, 58: 83–86.

Coty, V.F., 1976. Microorganisms of oil recovery: status of microbial oil recovery. In: *Proceedings, 1976 Engineering Foundation Conference, on The Role of Microorganisms in the Recovery of Oil, Tidewater Inn, Easton, Md*. NSF/RA-770201: 77–81.

Davis, J.B., 1967. *Petroleum Microbiology*. Elsevier, Amsterdam.

Donaldson, E.C., 1982. Microbial enhancement of oil recovery. *BETC Q. Progr. Rev.*, 31 (April–June).

George, A.A. and Chen, J.S., 1983. Acidic conditions are not obligatory for onset of butanol formation by *Clostridium beijerinckii* (Synonym *C. butylicum*). *Appl. Environ. Microbiol.*, 46(2): 321–327.

Grula, E.A., Russell, H.H., Bryant, D., Kenaga, M. and Hart, M., 1983. Isolation and screening of clostridia for possible use in microbially enhanced oil recovery. In: E.C. Donaldson and J.B. Clark (Editors), *Proceedings, 1982 International Conference on Microbial Enhancement of Oil Recovery*. U.S. Department of Energy, NTIS, Springfield, Va., pp. 43–47.

Hitzman, D.O., 1962, Microbiological secondary recovery. *U.S. Patent*, No. 3,032,472.

Hitzman, D.O., 1983. Petroleum microbiology and the history of its role in enhanced oil recovery. In: E.C. Donaldson and J.B. Clark (Editors), *Proceedings, 1982 International Conference on Microbial Enhancement of Oil Recovery*, U.S. Department of Energy, NTIS, Springfield, Va., pp. 102–218.

Ivanov, M.V. and Belyaev, S.S., 1983. Microbial activity in waterflooded oil fields, and its possible regulation. In: E.C. Donaldson and J.B. Clark (Editors), *Proceedings, 1982 International Conference on Microbial Enhancement of Oil Recovery*. U.S. Department of Energy, NTIS, Springfield, Va., pp. 48–57.

Jack, T.R., Thompson, B.G. and Diblasio, E., 1983. The potential for use of microbes in the production of heavy oil. In: E.C. Donaldson an J.B. Clark (Editors), *Proceedings, 1982 International Conference on Microbial Enhancement of Oil Recovery*. U.S. Department of Energy, NTIS, Springfield, Va., pp. 88–93.

Johnson, A.C., 1979. Microbial oil release technique for enhanced recovery. *Conf. on Microbial Processes Useful in Enhanced Oil Recovery. Final Rep.*, pp. 30–34.

Lazar, I., 1983. Microbial enhancement of oil recovery in Romania. In: E.C. Donaldson and J.B. Clark (Editors), *Proceedings, 1982 International Conference on Microbial Enhancement of Oil Recovery*. U.S. Department of Energy, NTIS, Springfield, Va., pp. 140–148.

Pines, O., Bayer, E.A. and Gutnick, D.L., 1983. Localization of emulsant-like polymers associated with the cell surface of *Acinetobacter calcoaceticus*. *J. Bacteriol.*, 154: 893–905.

Postgate, J.R., 1979. *The Sulphate-Reducing Bacteria*. Cambridge University Press, Cambridge.

Yarbrough, H.F. and Coty, V.F., 1983. Microbially enhanced oil recovery from the Upper Cretaceous Nacatoch Formation, Union County, Arkansas. In: E.C. Donaldson and J.B. Clark (Editors), *Proceedings, 1982 International Conference on Microbial Enhancement of Oil Recovery*. U.S. Department of Energy, NTIS, Springfield, Va., pp. 149–153.

Zhang, Z. and Qin, T., 1983. A survey of research on the application of microbial techniques to the petroleum production in China. In: E.C. Donaldson and J.B. Clark (Editors), *Proceedings, 1982 International Conference on Microbial Enhancement of Oil Recovery*. U.S. Department of Energy, NTIS, Springfield, Va., pp. 135–139.

ZoBell, C.E., 1946. Bacteriological process for treatment of fluid-bearing earth formations. *U.S. Patent*, No. 2,413,278.

ZoBell, C.E., 1947a. Bacterial release of oil from sedimentary materials. *Oil Gas J.*, 46: 62–65.

ZoBell, C.E., 1947b. Bacterial release of oil from oil-bearing materials. *World Oil*, 126: 36–47; 127: 35–41.

Chapter 7

MICROBIAL PLUGGING IN ENHANCED OIL RECOVERY

T.R. JACK, J. SHAW, N. WARDLAW and J.W. COSTERTON

INTRODUCTION

Although bacteria have long been recognized as agents of formation damage by those involved in the production of oil (Cerini et al., 1946; Beck, 1947), the importance of bacterial plugging in waterflooding operations is generally underrated. This is due partly to the fact that early laboratory investigations concentrated on the plugging effects of suspensions of dead cells (Hart et al., 1960; Kalish et al., 1964; Raleigh and Flock, 1965) and failed to systematically evaluate the more damaging effects resulting from the metabolism of the viable microorganisms found in the field.

In this chapter, the writers begin with an exploration of the microbial plugging problems associated with conventional waterflooding operations and evaluate the mechanisms responsible. Then, the writers define the characteristics required of bacterial MEOR cultures for their successful injection and examine the microbial problems anticipated during other EOR operations such as polymer flooding.

Throughout, considerations are confined to the wellbore region and do not include problems associated with the subsequent dispersion of bacteria through the reservoir, this being the subject of active study elsewhere (Jang et al., 1983a, b; see Chapter 8, this volume).

BACTERIA IN WATERFLOODING INJECTION WATERS

In the secondary production of oil, water is injected under pressure into the producing formation via designated injection wells in such a way that oil underground is displaced through the formation to production wells. The quality of water used in this operation varies from field to field and may include fresh water from wells or surface sources, produced water or, in some locations, sea water. In some instances, poor water quality results in plugging of the injection well at the formation face. This loss of permeability in the wellbore region results in a loss of "injectivity" (a decline in the rate of injection), which can seriously jeopardize the overall production of oil. This problem is not confined to waterflooding but is of general importance in all types of enhanced oil production (polymer flooding, alkaline flooding, micellar or microemulsion flooding, etc.), which rely on an injected "pusher" phase to displace the oil from the reservoir.

TABLE 7-I

Microorganisms found in 267 samples of wellhead injection waters in Alberta (based on data from Myers and Slabyj, 1962)

Type of microorganism	Percentage of samples
Molds	14
Yeasts	1
Algae	0.4
Diatoms	2
Actinomycetes	1
Slime-forming bacteria	30
Sulfate-reducing bacteria	38
Iron bacteria	14
Aerobes (over 10^5/ml) [a]	29
Anaerobes (over 10^2/ml)	16

[a] The arbitrary maximum permissible for injection water quality (Myers and Slabyj, 1962)

Early investigations of injectivity losses in waterflooding (Cerini et al., 1946; Beck, 1947) concluded that bacteria or bacterially produced organic substances could play a role in the plugging process.

Myers and Slabyj (1962) examined 471 samples of injection water in Alberta, 267 of which came directly from the wellhead at injection wells. The authors concluded that 67% of the wellhead samples contained high levels of deleterious bacteria and that these bacteria could be present even in systems where the injection water had been drawn from freshwater wells. Some of the types of microorganisms found are summarized in Table 7-I. Although the types of microorganisms listed may seem somewhat arbitrary, molds, algae, iron bacteria, slime-forming bacteria, and sulfate-reducing bacteria have all been cited as microbes potentially able to cause plugging (Lada, 1959; Trautenberg, 1965). Whereas molds and algae are cited as undesirable on the basis of the relatively large size of their cells, the bacteria tabulated have been regarded as potential plugging agents as a consequence of their ability to proliferate rapidly and produce slimes and precipitates.

The sulfate-reducing bacteria found in 38% of the wellhead samples (Table 7-I) are a group of obligate anaerobes, which reduce sulfate to sulfide in a manner analogous to that in which aerobes reduce oxygen to water during aerobic metabolism. The production of copious quantities of sulfide results in a health hazard from H_2S release, problems of corrosion, and potential plugging through the precipitation of iron sulfide minerals (principally mackinawite) in situ (Davis, 1967; Lada, 1959). Ashirov and Szonova (1962) also proposed that reduction of sulfate can lead to dissolution of $CaSO_4$ minerals (presumably gypsum or anhydrite) with a consequent precipitation of the liberated Ca^{2+} as calcite, $CaCO_3$. This they demonstrated in the laboratory and inferred as the mechanism whereby secondary calcite precipitation had sealed the oil–water zone in certain reservoirs. In one case, such a calcite barrier sealed off the oil-bearing production zone from an underlying aquifer despite a pressure differential of 58 atm (Ashirov and Szonova, 1962).

The iron bacteria found in 14% of the wellhead samples (Table 7-I) can potentially cause plugging through the precipitation of iron oxides often in association with microbial slime (Lada, 1959; Trautenberg, 1965; Davis, 1967).

The slime-forming bacteria found in 39% of the wellhead samples (Table 7-I) have been implicated in the plugging of sand packs (McCalla, 1950; Mitchell and Nevo, 1964; Gupta and Swartzendruber, 1962) and soils (Allison, 1947) as discussed later. Hensel et al. (1981) suggested that injection waters should contain less than 10 slime-forming bacteria per milliliter to avoid plugging problems.

Bacteria are present in all injection waters and are introduced into the reservoir itself during waterflooding. Moreover, the above studies indicate that some of the most worrisome groups of bacteria—those able to produce precipitates and slime through their metabolic activity—are among the bacteria commonly present.

PREVIOUS STUDIES OF BACTERIAL PLUGGING

Despite a general understanding in the oil industry that precipitates and slime generated by actively growing microorganisms can be important in formation plugging (Lada, 1959; Trautenberg, 1965; Davis, 1967), the study of plugging mechanisms using live bacteria has been limited.

Plugging by dead bacteria

Early studies based on the injection of dead bacterial cell suspensions through sandstone cores have been reviewed in detail by Davis (1967). In Berea Sandstone having a permeability of 48–117 mD, the mean pore size of 7–8 μm in diameter is comparable to bacterial cell dimensions (typically 1–3 μm for single cells). Since dead cells cannot produce copious slimes, precipitates, etc., the plugging observed is due to the cells themselves acting as particles.

The general nature of particulate plugging has been investigated by Gruesbeck and Collins (1982). In their experiments, suspensions of fine particles (glass beads) of known diameter were injected into sand packs in which the average "throat" size of the interconnected pores had been determined. It was found that where the ratio of throat size to particle diameter exceeded 13, no significant plugging was observed, although particles could adsorb as a layer on the surface of the sand grains throughout the sand column (Fig. 7-1A). At ratios from 4 to 6, both surface adsorption and plugging due to a "log jam" effect in the pore throats were observed (Fig. 7-1B). At ratios below 2.6, the particles formed a filter cake on the face of the sand packs (Fig. 7-1C) (Gruesbeck and Collins, 1979). In plugging studies using dead bacteria and Berea Sandstone cores, plugging and filter cake formation were observed.

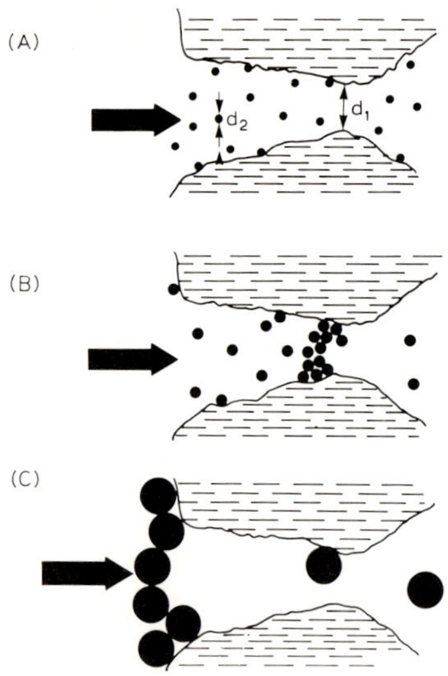

Fig. 7-1. Particulate plugging can be categorized into three distinct regimes depending on the ratio of throat size (d_1) of the interconnected pore space to the particle diameter (d_2): (A) surface adsorption on rock surfaces causing minor permeability reduction, $d_1/d_2 \geqslant 13$; (B) plugging in pore throats within the core as well as surface adsorption effects, $d_1/d_2 = 4$–6, and (C) formation of a filter cake on the top face of the core in addition to plugging and surface adsorption effects, $d_1/d_2 \leqslant 2.6$ (after Gruesbeck and Collins, 1979).

The rate of plugging depends on the density of the bacterial cell suspension, but is relatively insensitive to the initial permeabilities of Berea Sandstone cores over the limited range tested, i.e., 123 to 310 mD (Hart et al., 1960). Whereas bacteria do penetrate the entire length of the cores, the major accumulation of cells and consequent plugging is confined to the first few centimeters of core length (Hart et al., 1960; Kalish et al., 1964). Lower-concentration cell suspensions form plugs to a greater depth than do high-concentration cell suspensions (Hart et al., 1960). Kalish et al. (1964) found that the plugging observed depended in a predictable manner on the size, shape, and aggregation of the bacterial cells. Thus long chains of rod-shaped cells (each 1 μm in diameter $\times$ 5 μm in length) were more efficient plugging agents than small aggregates of spherical cells (each 1 μm in diameter), which in turn were more efficient than singly dispersed cells (1 μm diameter $\times$ 2.5 μm length). According to Kalish et al. (1964), permeability was never reduced to zero . In the worst cases, permeability reductions of 90% were reported for low-permeability cores (30 mD) and 95% for high-permeability cores (300 mD), presumably due to the fact that the "log jams" of the cells formed in the core were themselves permeable (Kalish et al., 1964). For cores which were partially plugged, Darcy's law did not apply because increases in injection pressure were not linearly related to flow rate (Kalish et al., 1964). Possibly, increasing pressure dislodged the cells plugging capillaries, perhaps by shearing some of the cells. Attempts to quantitatively relate plugging of cores to the physical characteristics of various types of rocks were not very successful. Although the depth of plug formed by the dead bacterial cells was found to be a significant function of pore geometry, no single reliable correlation was found for the rate of plugging (Raleigh and Flock, 1965).

The general picture emerging from these studies is that dead cells alone do cause plugging in sandstone cores having initial permeabilities similar to many oil-bearing formations. Plugging seems to be initiated at the inlet end of the cores where clumps of cells fortuitously present in suspension are preferentially retained. Intact cells were always observed in the effluents from the core.

Unfortunately, the results obtained from studying the plugging effects of dead bacteria are not particularly relevant to the problem experienced in the oilfield. Whereas it is true that Hart et al. (1960) found that the plugging effects of live and dead cells of a pure strain of *Bacillus subtilis* were similar, the sweeping generalization of Kalish et al. (1964) that "there is no reason to suppose that results with live cells would be different" is not valid. As discussed in the next section, the most severe plugging by bacterial cultures is not due to the bacterial cells themselves but rather to the formation of metabolic products, particularly polysaccharides. Widespread acceptance of Kalish's results as being applicable to the field has led to a serious underestimation of the potency and prevalence of microbial plugging effects.

Plugging by live bacteria

In 1947, L.E. Allison of the U.S. Department of Agriculture reported that the permeability of submerged soils was reduced over time by the viable microorganisms. He speculated that the observed sealing effect in organic-rich loams was due entirely to microbial activity which clogged the soil pores "with the products of growth, cells, slimes or polysaccharides" (Allison, 1947). McCalla (1950) reported that the plugging observed during prolonged water percolation through soils depended on the organic material available. Thus, in Peorian loess, which contains only 0.5% oxidizable materials, percolation rates remained almost constant over fifty days. In the absence of additional organic nutrients, the loess supported only a meagre microbial population (McCalla, 1950). The addition of a 1 or 2% sucrose solution to any of the soils tested enriched the microbial population in the soil and increased the rate and extent of permeability loss observed on prolonged percolation. This was particularly dramatic in the case of the organic-deficient Peorian loess (Fig. 7-2). It was also noted that the addition of biocides able to prevent microbial growth prevented plugging (McCalla, 1950).

Similar studies have been carried out using columns packed with coarse sand rather than soil. The relatively high permeability of these sand packs suggests that the average throat size of the pores in the sand pack will be large enough to allow

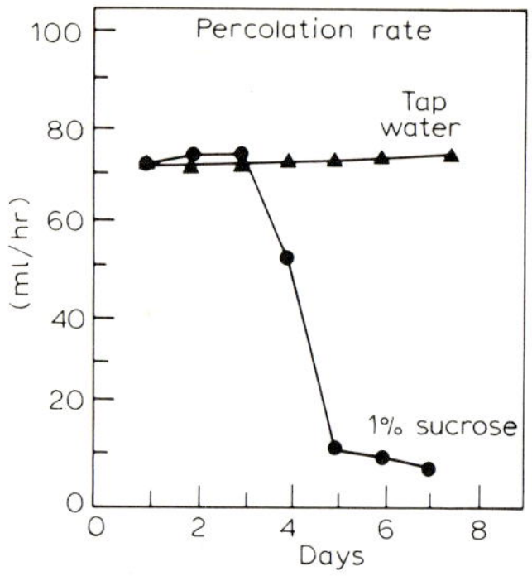

Fig. 7-2. Effect of adding sucrose solution to a percolating column packed with organic-deficient Peorian loess (after McCalla, 1950).

passage of singly dispersed bacterial cells as predicted by Gruesbeck and Collins (1979) (Fig. 7-1).

Nevertheless, Gupta and Swartzendruber (1962) found that after prolonged flow of boiled deionized water through a septic quartz sand column (Ottawa sand, 50/70 mesh), substantial plugging did occur. No such decrease in permeability was observed in the presence of a biocide (0.1% phenol). Measurements of the pressure drop across various sections of the column indicated that plugging occurred in the first 10% of the column length (the first 1.5 cm) which correlated with the presence of a large bacterial population in this zone. Indeed for a series of experiments, an excellent correlation was found between the percentage reduction in the permeability of the sand pack and the number of bacteria actually present in the sand (Fig. 7-3). Further calculation showed that bacterial cells could not reasonably account for the plugging observed, because the cells themselves occupied less than 0.008% of the void volume available. The authors, therefore, concluded that plugging was due primarily to products of bacterial metabolism (polysaccharides, slimes, precipitates, etc.) and not to the cells themselves (Gupta and Swartzendruber, 1962). Since the sand itself provided no organic nutrients, trace organics present in the deionized water (perhaps originally leached from the ion exchange resins) must have provided the nutrients necessary for microbial growth. This is consistent with the observation that microbial growth and plugging occurred on the front face of the sand pack where fresh nutrients would constantly be introduced by the incoming water. The developing microbial population in this zone would deplete nutrients from the water and hence inhibit growth and consequent plugging downstream in the rest of the sand column.

The importance of bacterial products as agents of plugging was confirmed by Mitchell and Nevo (1964). These researchers used lysimeters packed with coarse sand having less than 0.03% organic content. The effects of adding organic nutrients (casein), Fe^{2+}, and S^{2-} to the top layer of sand were studied with respect to the

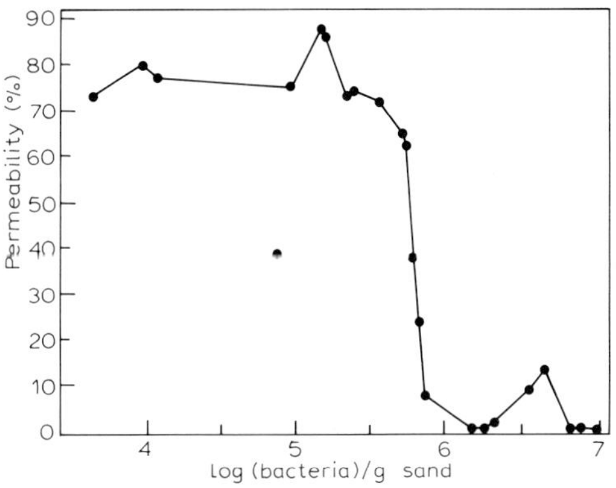

Fig. 7-3. Relationship between loss of permeability and the number of bacteria present in a sand-packed column after 25 days percolation with deionized water (after Gupta and Swartzendruber, 1962).

extent and type of plugging observed. A strong correlation was found between plugging and the concentration of polysaccharides formed by viable bacteria in the sand (Fig. 7-4). Casein addition enhanced polysaccharide formation, particularly by the *Flavobacteria*, and consequent plugging resulted regardless of the addition of Fe^{2+} and S^{2-} or of the formation of black iron sulfides. Thus the intentional chemical precipitation of iron sulfides in columns did not result in significant permeability loss and in cases where black iron sulfides were formed as a result of biological activity under reduced (anaerobic or microaerophilic) conditions, the plugging observed was still attributable to the formation of polysaccharides. Apparently, the precipitated iron sulfide particles by themselves were too fine to cause significant permeability loss in the coarse sand packs employed.

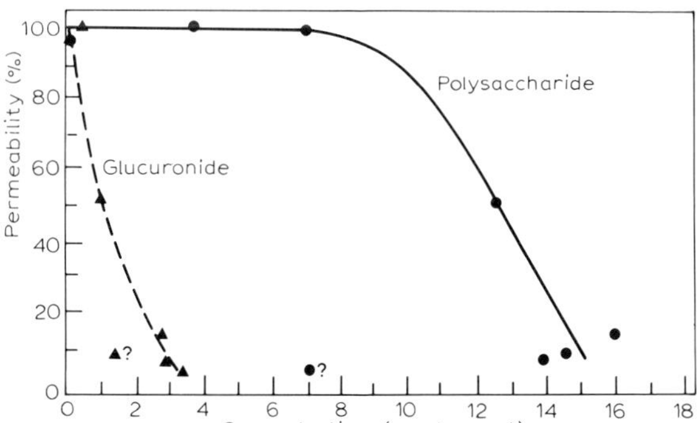

Fig. 7-4. Correlation of total bacterial polysaccharides and component glucuronide with permeability loss for tap water percolation through a coarse sand column amended with organic nutrients (after Mitchell and Nevo, 1964).

Mitchell and Nevo (1964) also suggested that the stability of the polysaccharide plug formed was related to its biodegradability. Formation of more complex polysaccharides having a uronic acid component, as promoted by organic nutrients having an appropriate C/N ratio, generated a more stable plug. The correlation of plugging with total polysaccharide present or with the glucuronic acid component was significant in either case (Fig. 7-4).

In general, coarse sand packs under continual waterflood are effectively sealed by the formation of microbial polysaccharides and biomass. The location and degree of microbial activity depends on the availability of organic nutrients. Where these are brought into the sand by the incoming flood water, microbial growth, activity, and consequent plugging is greatest in the front face of the sand pack. Since coarse sand packs have an initial permeability higher than that found in most oil-bearing formations, bacterial plugging through exopolymer formation is anticipated to be a very effective cause of plugging in waterflood operations in the oilfield. By analogy, the microbial plugging of an injection well is expected to occur in the zone immediately adjacent to the wellbore, where the microorganisms are constantly receiving fresh nutrients from the incoming water.

This does not imply that microorganisms introduced into the reservoir with the flood water will be confined to the region of the wellbore of the injection well. On the contrary, Myers and McCready (1966) showed that viable bacteria could penetrate even very tight limestones. Under a pressure differential of less than two atmospheres, cell suspensions of *Serratia marcescens* (spherical cocci of about 1 μm in diameter), were induced to traverse 14 inches of Berea Sandstone (100–300 mD permeability) in 84 hours and 3 inches of Devonian limestone (1.7 mD) or late Mesozoic or Mississippian limestone (less than 0.1 mD) in longer times. Even in the absence of a pressure gradient, the bacteria were found to permeate entire Berea Sandstone cores in all cases tested. It seems likely that bacteria carried into the reservoir in injection waters will disperse over significant volumes of the reservoir even under low pressure differentials. The general inability of bacteria to metabolize oil quickly under anaerobic conditions (Davis, 1967; Ward and Brock, 1978), however, will limit their subsequent growth and activity away from the wellbore region (Anonymous, 1972). In the wellbore region of the injection well, trace organics and oxygen in the incoming injection water will support a more prolific and productive microbial population.

Surface growth by bacteria

In the injection well of a waterflood, some bacteria can attach themselves to the rock surfaces initially encountered by the incoming injection water. From extensive studies carried out in industrial applications as well as in the areas of medicine and ecology (Costerton et al., 1978, 1981; Geesey et al., 1978; Costerton and Geesey, 1979a, b; Bitton and Marshall, 1980; Costerton, 1981) it is evident that surface growth by bacteria to form biofilms occurs ubiquitously, even on smooth surfaces subject to relatively high shear forces (McCoy et al., 1981). This is often the preferred growth habit in nature (Geesey et al., 1978; Costerton and Geesey, 1979a, b; Geesey and Costerton, 1979).

Whereas the mechanisms of attachment of bacterial cells to surfaces are numerous (Bitton and Marshall, 1980), a key feature of non-specific general attachment is the production of extracellular "slime" rich in polysaccharides, referred to as the glycocalyx. The glycocalyx forms a sheath about the bacteria, which mediates surface adherence and offers protection against many biocides (Costerton, 1981; Ruseska et al., 1982) and some natural enemies. It also promotes growth by sequestering organic and inorganic nutrients from aqueous solution and retaining nutrients released by the cells enclosed within the biofilm (Costerton et al., 1981).

Given these ecological advantages, it seems probable that the viable bacteria responsible for plugging the face of the laboratory sand packs discussed above or the wellbore region of the injection well in a waterflood are immobilized in a polysaccharide-rich glycocalyx. This is consistent with the laboratory observations presented in Figs. 7-3 and 7-4.

MECHANISMS OF MICROBIAL PLUGGING IN THE FIELD

The above review and discussion leads to the conclusion that there are two general mechanisms of microbial plugging likely to occur in the wellbore region.

Particulate plugging

It is apparent from studies using dead cell suspensions and Berea Sandstone cores that bacterial cells alone can cause permeability reduction by plugging and filter cake formation. The importance of particulate plugging, however, depends on the ratio of the average throat size of the pores in the formation to the size of the bacterial cells or their aggregates (chains, clusters, etc.) (Gruesbeck and Collins, 1979; Fig. 7-1). Thus, in very tight formations even singly dispersed bacterial cells could cause an injectivity problem, whereas in more permeable formations, only larger aggregates of cells would be of concern.

Plugging due to biofilm formation

The buildup of biofilms in the wellbore region by viable microorganisms able to colonize the rock surfaces constitutes a second general plugging mechanism. In contrast to the particulate plugging mechanism, the formation of an extensive glycocalyx can bridge the pore throats of the formation face effectively reducing its permeability to zero. A simplified sequence of events leading to biofilm formation is illustrated in Fig. 7-5.

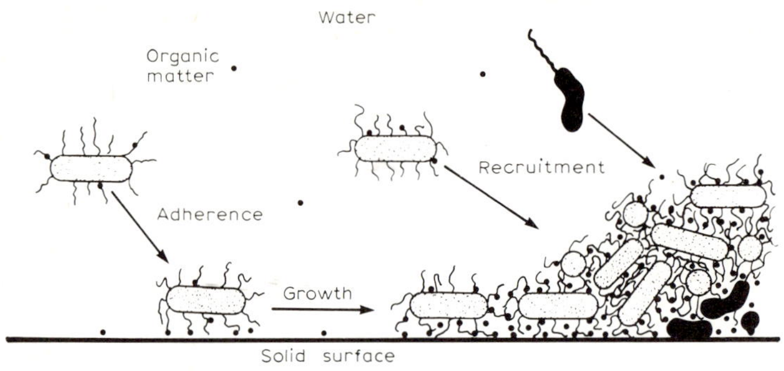

Fig. 7-5. Schematic scheme for the colonization of an inert solid surface by adherent bacteria to generate a biofilm in which the glycocalyx surrounding the cells holds the bacteria on the surface, sequesters organic nutrients and recruits other bacteria from the planktonic population.

In an actual injection well, the population of naturally adherent bacteria would be supplemented initially by the retention of other viable and dead bacteria either as clumps or individually according to the parameters illustrated in Fig. 7-1. As the biofilm matures, the efficiency of recruiting bacteria into the plug from the incoming injection water would increase, because the affinity of microbial cells for the exopolymers of the glycocalyx will be generally greater than that for the bare inorganic surfaces (Tosteson et al., 1982).

DEMONSTRATIONS OF MECHANISMS OF MICROBIAL PLUGGING

The mechanisms of microbial plugging have been independently demonstrated by the writers in their laboratories in a model core system in which the average pore throat size is sufficiently large (33 μm) to ensure that particulate type plugging will be relatively insignificant for suspensions of singly-dispersed bacteria.

Model core system

The model core system consists of (1) a reservoir for the cell suspension, which can be pressurized to 3.5 psi by nitrogen or mixed gases; (2) a core holder with a pressure gauge at the inlet end; and (3) a volumetric collection device for the effluent. Plugging is measured by monitoring decreases in the flow rate under constant pressure (Shaw, 1982).

The core itself consists of uniform glass beads, 90 μm in diameter, fused into a coherent cylinder, 1 cm in diameter and 5 cm in length. The core has 38% porosity, 6.5 D permeability, and 33 μm mean pore throat size (Shaw, 1982).

Plugging studies were routinely carried out using a previously sterilized apparatus through which various bacterial cultures were injected.

Where sterile distilled water was employed as the flooding medium, no loss in permeability was evident over 18 hours and 13,400 pore volumes injection (Shaw, 1982).

Cores were routinely sectioned after each run and examined by scanning electron microscopy.

Particulate plugging in the model core system

Given a mean pore throat size of 33 μm and the criteria of Fig. 7-1, filter cake formation in the model core system would be anticipated only where large aggregates of bacterial cells were injected into the core.

This was indeed found to be the case using a wild isolate, which was an anaerobic rod-shaped bacterium 592-3, measuring 1 μm in diameter and about 2 μm in length. On two occasions these bacteria were grown in a saline nutrient broth (DIFCO) medium, supplemented with sucrose on one occasion and a mixture of fructose and glucose on the other.

In the former case, the growth habit was single dispersed cells, whereas in the latter, growth occurred in extended chains of up to 25 cells. As was observed by Kalish et al. (1964), the aggregates of cells were more efficient plugging agents than the singly-dispersed cells (Fig. 7-6). Inspection of the cores using a scanning electron microscope showed that cells had accumulated in both cores, but that the singly-dispersed cells had accumulated principally in the face of core near the inlet end, whereas the extensive chains had formed a filter cake on top of the core face (Fig. 7-7). The extracellular polymer formation was not observed in either case (Fig. 7-8). The short duration of the runs allowed only a doubling of cell numbers in the core apparatus.

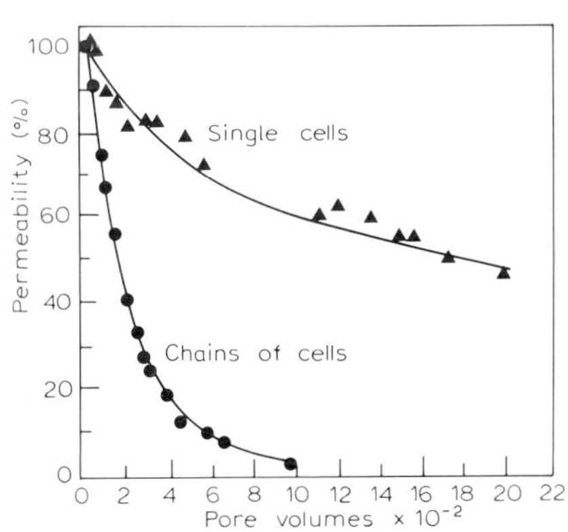

Fig. 7-6. The effect of growth habit (singly dispersed cells, or extensive chains of cells) on the plugging observed for wild anaerobic isolate 592-3 in the model core system. In both cases, actively growing cells were passed through the core at approximately 3×10^8 cells/ml. The actual elapsed time (about 4 hours) was short enough to preclude significant growth of the culture in the core.

Comparison of the distributions of chain lengths in the original cell suspension and in the core below the filter cake (Fig. 7-9) showed that the filter cake had selectively retained chains containing five or more cells. This corresponds to a mean pore throat size to chain length ratio of about 3.3 in agreement with the prediction

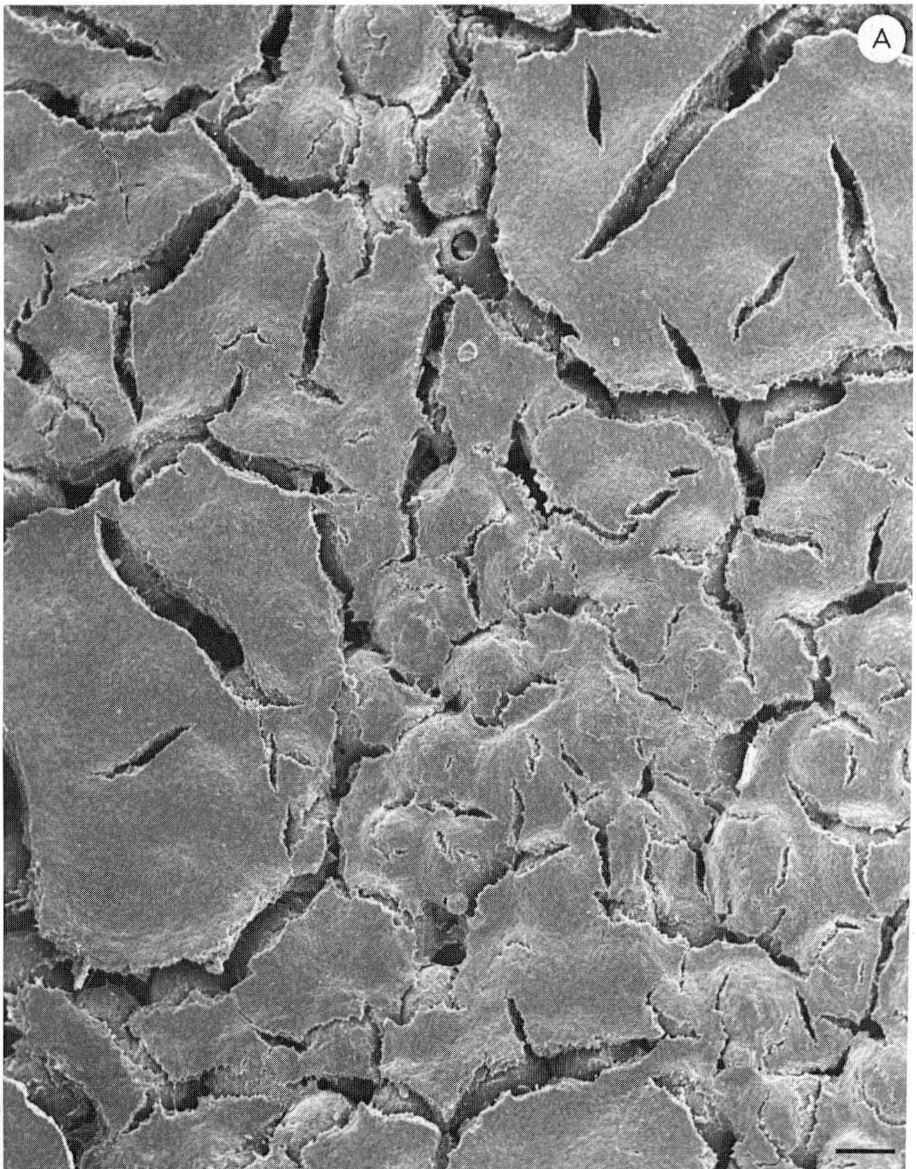

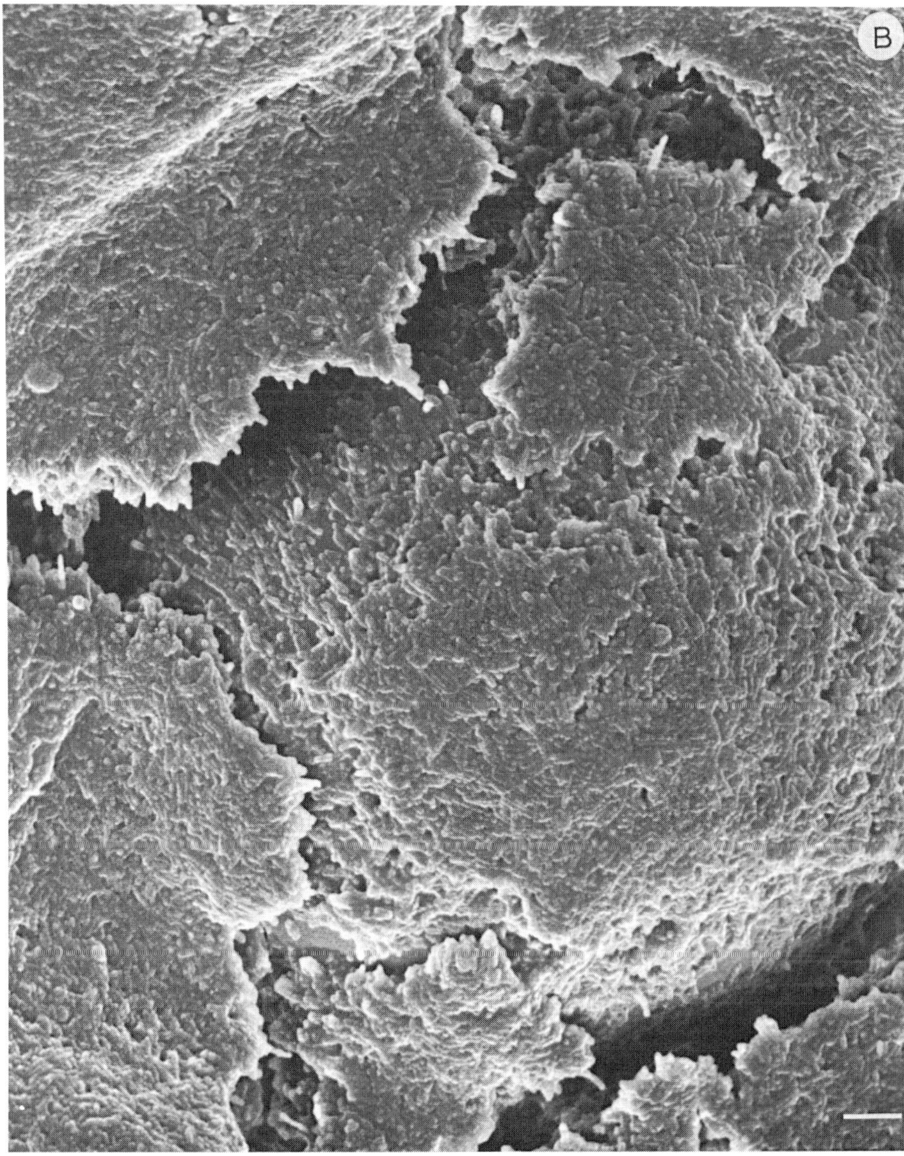

Fig. 7-7. A. Low-magnification scanning electron microscope image of the thick filter cake on top of the model core (size bar, 50 μm). B. High-magnification image of the filter cake showing it to be a collection of chains of cells (size bar, 5 μm).

that filter cakes will form at ratios of 2.6 or less (Fig. 7-1). In view of the fact that Gruesbeck and Collins (1979) worked with spherical glass beads and not chains of cells, and that the cell suspension represents a wide distribution of chain lengths,

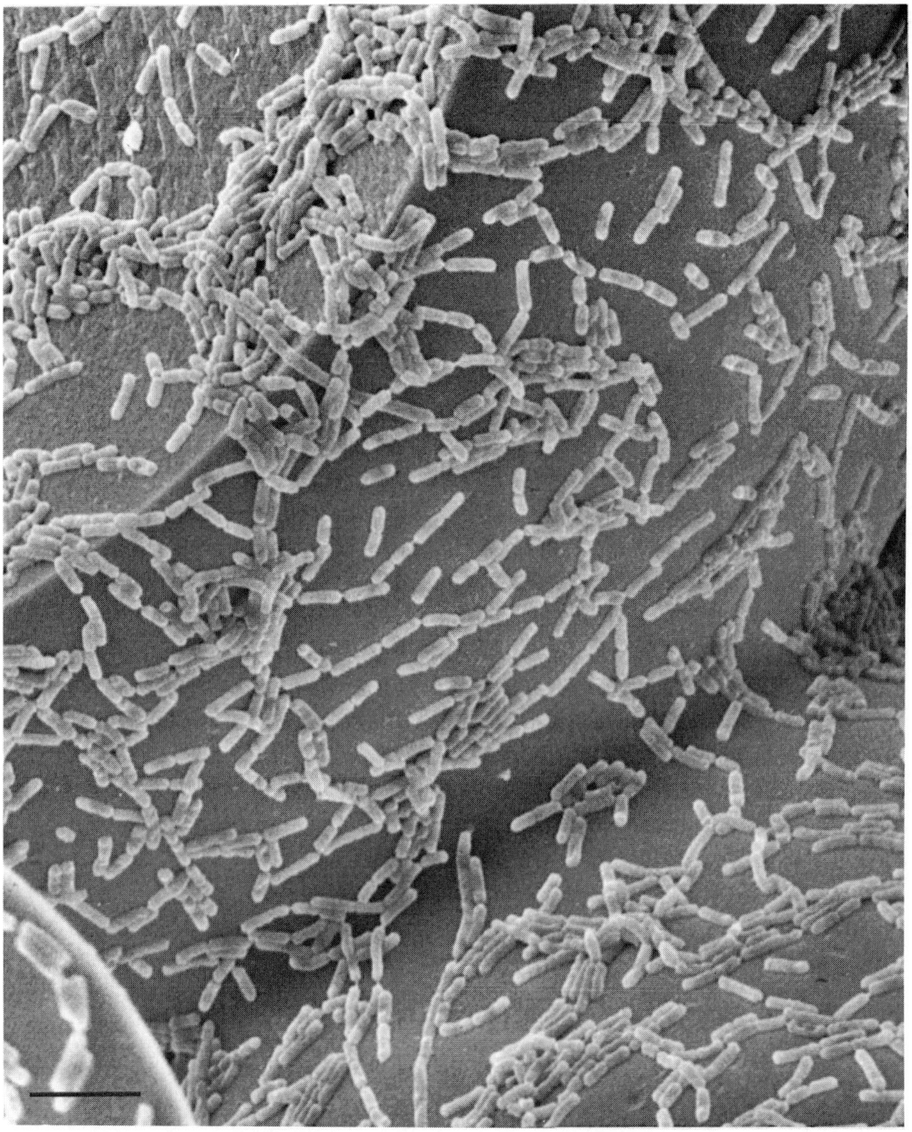

Fig. 7-8. High-magnification scanning electron microscope image of bacterial cells inside the model glass bead core below the filter cake. No exopolymer (slime) production is evident (size bar, 5 μm).

this is a surprisingly good agreement. In addition, complete loss of permeability was not observed (Fig. 7-6), which is consistent with earlier observations made using dead cell suspensions (Kalish et al., 1964). It appears then that even in this highly-permeable model core with pore throats larger than those found in most oil-bearing formations, bacterial aggregates of sufficient size can act as plugging agents by the particulate plugging mechanism.

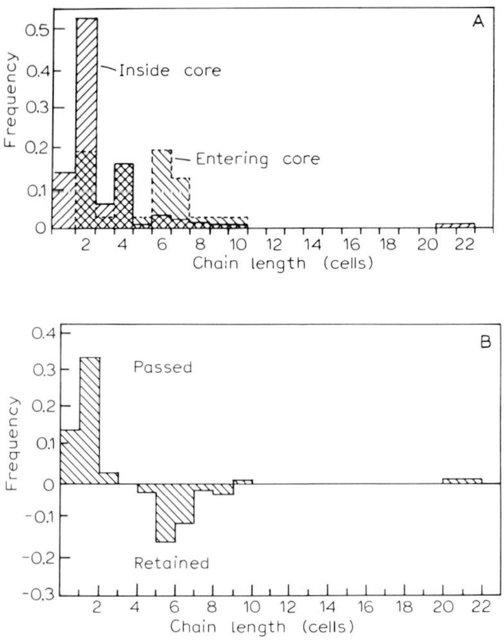

Fig. 7-9. A. The distribution of chain lengths above and below the filter cake formed on top of the model core by extensive chains of wild isolate, 592-3. B. The difference of the chain length distribution above and below the filter cake showing the selective retention of chains longer than five cells.

The very gradual plugging noted for the singly-dispersed cells (Fig. 7-6) may arise from clumps of cells fortuitously present in the cell suspension in accord with the criteria of Fig. 7-1. Plugging (not filter cake formation) was noted in the core face.

In both cases, the effluent from the core contained a large population of bacterial cells, only a small fraction of the total suspension being retained in the plug.

Plugging by biofilm-forming bacteria

Plugging by a mixed population

Where sterile distilled water was used as the flooding medium, no plugging was observed (Fig. 7-10). However, for non-sterile distilled water taken directly from a distilled water tap in the University of Calgary distribution system, plugging did occur. Permeability was reduced to 26% of its initial value over 18 hours (7700 pore volumes) as shown in Fig. 7-10. Examination of the sectioned core showed the presence of a mixed microbial population within the core attached to the surface of the glass beads by filaments of extracellular polymer (Fig. 7-11). Since electron microscope samples must be dehydrated prior to inspection, these filaments are in fact the very shrunken vestiges of a highly hydrated polymer matrix, which might have contained 99% water in its original state. Originally then the polymer matrix

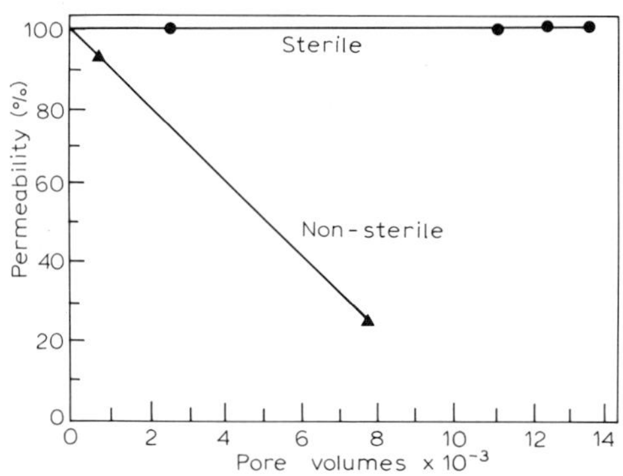

Fig. 7-10. Permeability reduction in the model glass bead core after 18 hours flooding with sterile and non-sterile distilled water (from Shaw, 1982).

would have largely filled the pore space available. This run gave the results anticipated for the mechanism illustrated in Fig. 7-5 and is analogous to a waterflooding operation using very clean, "nutrient-free" fresh water. The result also duplicates the observations of Gupta and Schwartzendruber (1962) using a non-sterile packed sand column and deionized water. Thus, microbial plugging due to "biofilm" formation can be expected as a universal phenomenon where biocides are not used. The rate and severity of plugging are, unfortunately, difficult to predict, being a function of the quantity and kind of viable bacteria and nutrients present in the injection water as well as the characteristics of the formation (temperature, pH, salinity, pore throat size distribution, etc.).

Dead and live bacteria

Two core-plugging runs based on dead and live cell suspensions of the *Pseudomonas* species isolated from biofilms on the rock surfaces of Marmot Creek, Kananaskis, Alberta, illustrate the efficiency and efficacy of viable biofilm-forming bacteria as plugging agents.

Viable rod-shaped cells of an exponentially growing culture in a citrate and mineral salts medium were found to rapidly and effectively plug the model core system. After 2500 pore volumes, the permeability was reduced to 0.3% of its initial value (Fig. 7-12). Bearing in mind that the plugging rate is proportional to the density of the cell suspension, this result is far more impressive than that noted for wild isolate 592-3 (Fig. 7-6) at a ten-fold higher cell density.

Inspection of the inlet surface of the core showed that the live microbes had sealed the core face to a depth of about 1 cm with a matrix of cells and extracellular slime (glycocalyx) (Fig. 7-13).

In contrast, a comparable culture killed by the addition of 5% formaldehyde proved far less effective and efficient (Fig. 7-12). Scanning electron microscope

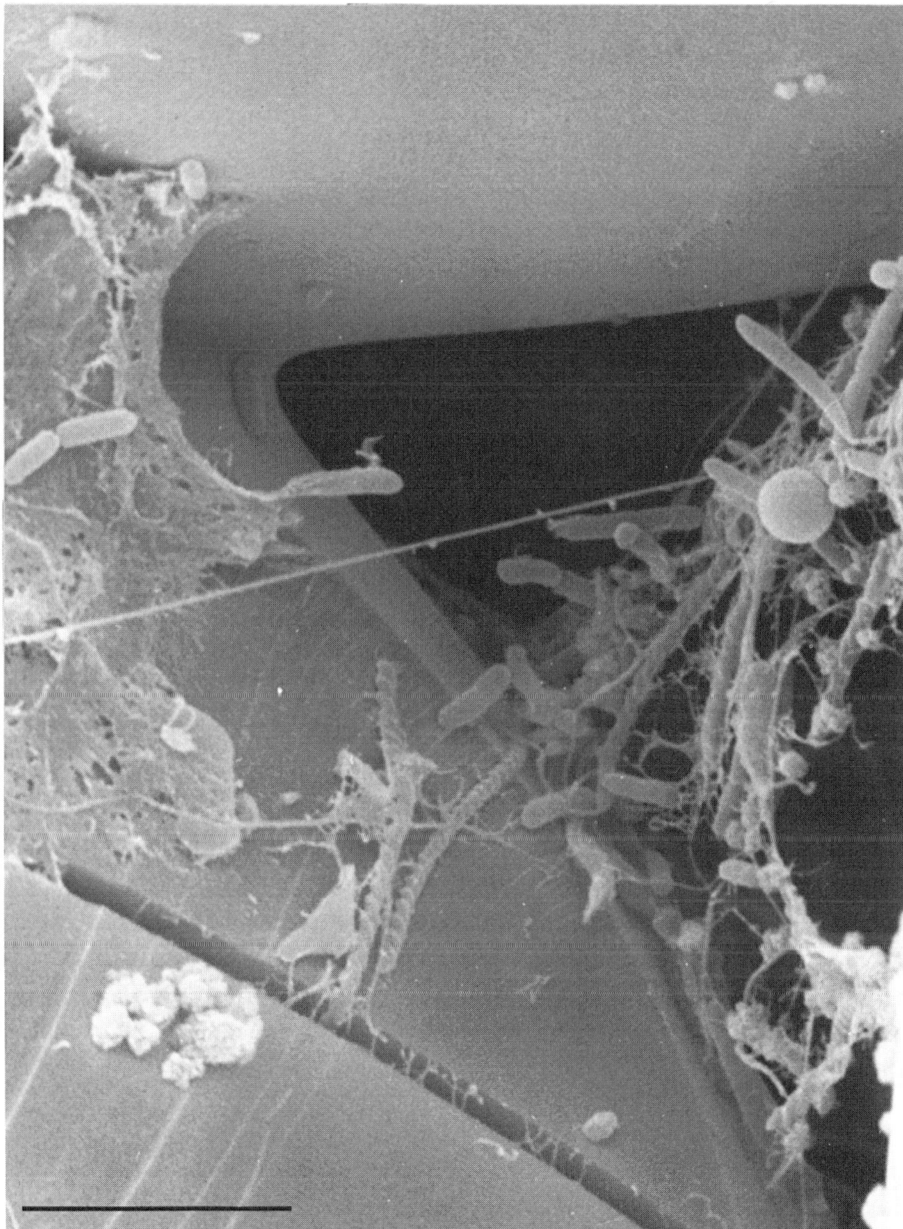

Fig. 7-11. High-magnification scanning electron microscope image of the mixed population of bacteria plugging the model core (from Shaw, 1982). Note especially the dehydrated and shrunken fibrils of exopolymers, probably polysaccharides, anchoring the cells to the glass bead surface. This is biofilm formation as anticipated in Fig. 7-5 (size bar, 5 μm).

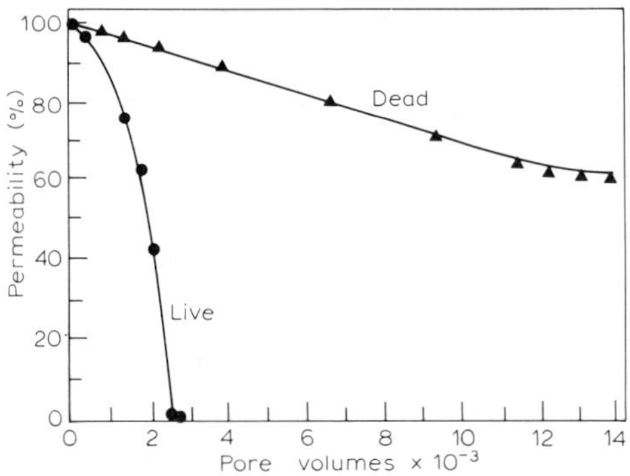

Fig. 7-12. Plugging by live and dead cell suspensions of a *Pseudomonas* species known to favor biofilm formation (from Shaw, 1982). Cell suspensions were $2-3\times10^7$ cells/ml and elapsed times were 23 and 29 hours for the live and dead cells, respectively.

inspection showed the core surface to be partially plugged with clumps of cells largely free of extracellular polymers (Fig. 7-14). Whereas it might be argued that the formaldehyde had affected the plugging result, this seems unlikely because the dead cell plugging curve (Fig. 7-12) matches that obtained for the polymer-free singly dispersed cells of wild isolate 592-3 (Fig. 7-6). Comparison is made on the basis of cumulative total cells injected rather than simply pore volumes of injected fluids. This supports the conclusion that the dead cells of the *Pseudomonas* are basically acting by the particulate plugging mechanism. Had longer study times been employed for 592-3 isolate, culture growth within the core might have given greater plugging for the live culture case.

Apparently, for this core system the ability of live cells to adhere and produce an extracellular polysaccharide-rich glycocalyx results in a more effective plugging than is observed for live cells. The latter do not produce extracellular polymer even where they grow as aggregates (592-3 chains, Fig. 7-6). It must be emphasized that in tighter formations, the plugging rate (efficiency) due to the particulate mechanism would be enhanced, but the ability of the biofilm to seal the formation face will remain unrivalled.

Extracellular polysaccharide

The importance of extracellular polymer production in microbial plugging has also been demonstrated using a culture not noted for biofilm formation. Wild isolate 18-1, a spherical coccus 1 μm in diameter, has been identified as a *Leuconostoc* species. This organism has no special affinity for the glass bead surface and produces a soluble polysaccharide, dextran, only from sucrose. Thus, comparable cultures can be grown anaerobically with polymer production on a medium containing sucrose or without polymer production on a medium containing fructose/glucose (1 : 1).

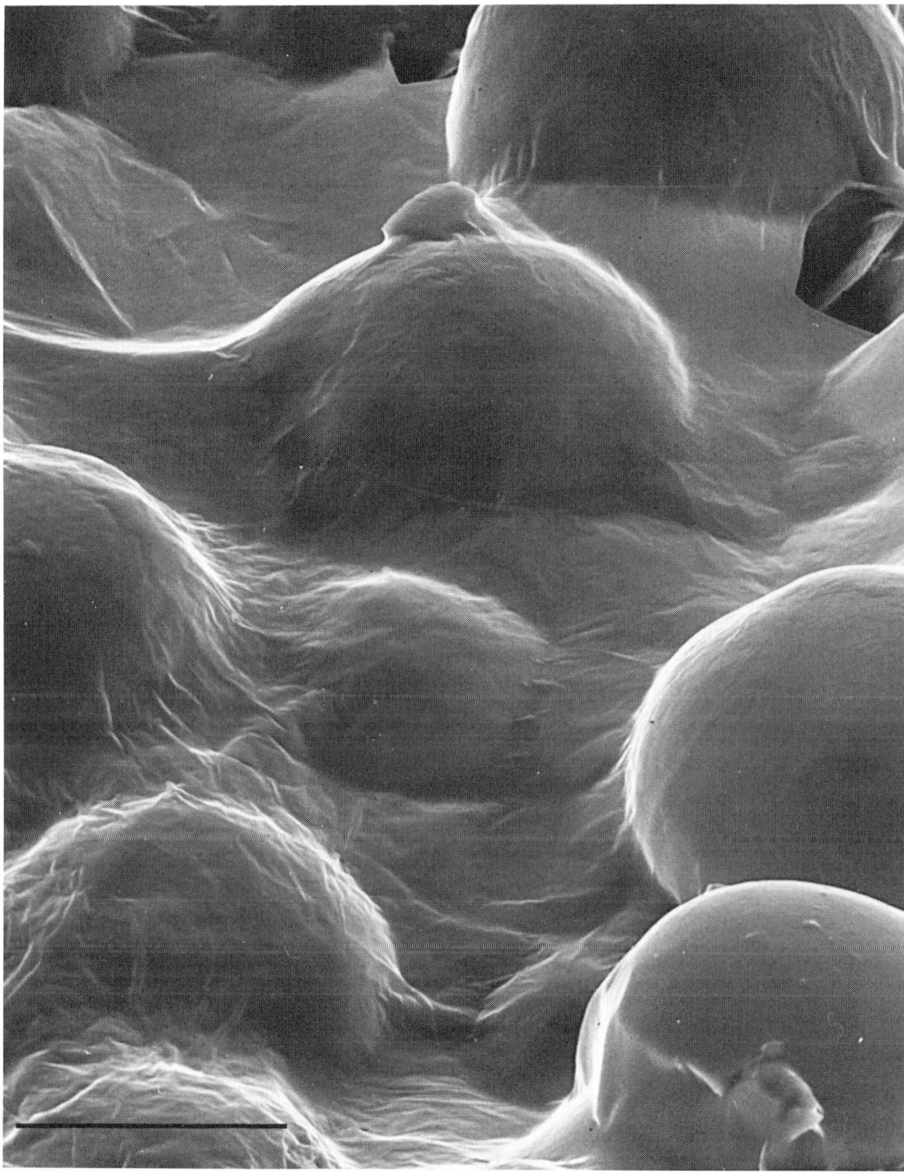

Fig. 7-13. Low-magnification scanning electron microscope image showing extensive polysaccharide (slime) production by the live cells of *Pseudomonas* (size bar, 50 μm) (from Shaw, 1982).

From Fig. 7-15, it is obvious that the production of evén a soluble exopolymer greatly enhanced the plugging ability of the culture. Inspection of the sectioned core showed the same result as was found for the *Pseudomonas* (Figs. 7-13 and 7-14): a plug in the top zone of the core consisted of cells in a matrix of slime for the

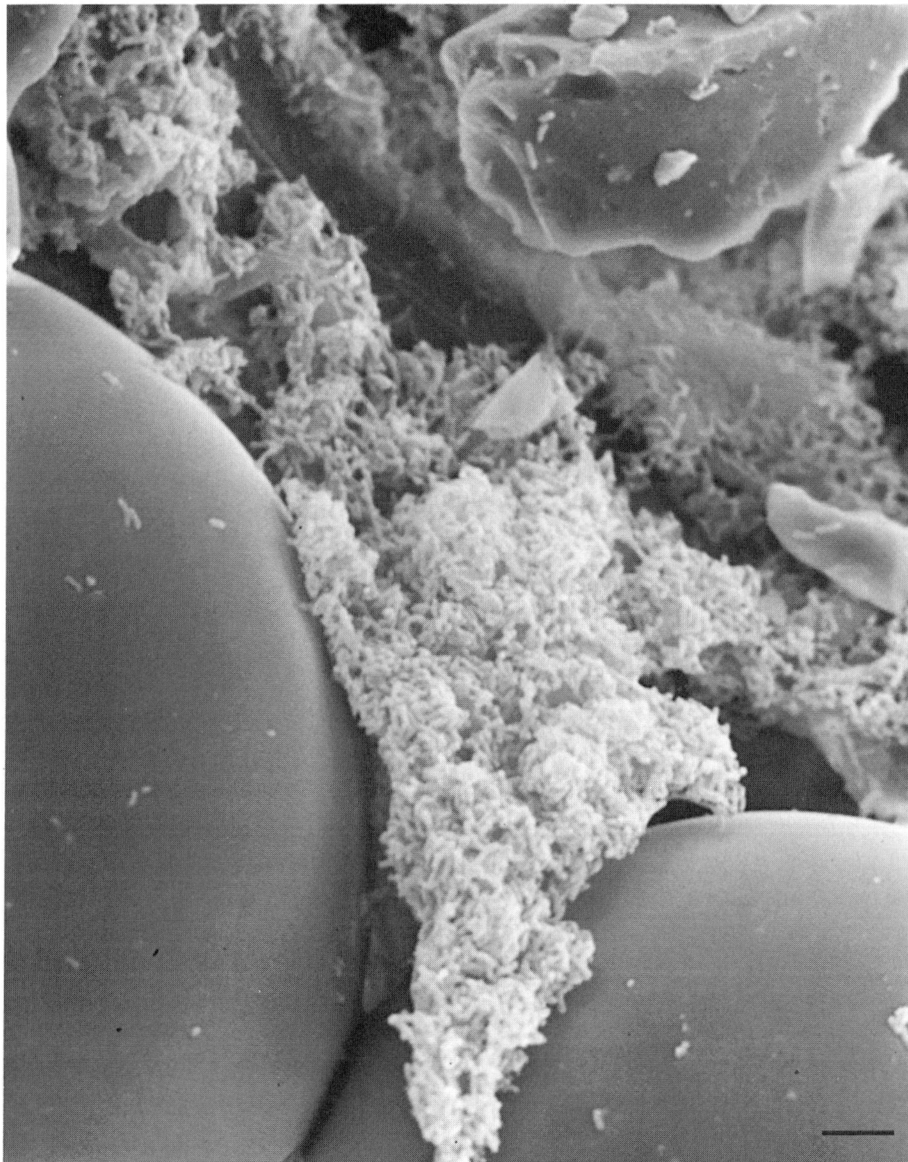

Fig. 7-14. High-magnification scanning electron microscope image of the clumps of dead cells of *Pseudomonas* which cause particulate plugging in the top zone of the model glass bead core. The absence of exopolymer production is evident in contrast to the live cell case (size bar, 5 μm) (from Shaw, 1982).

polymer-producing culture but only clumps of bare cells where polymer was not formed.

A further important observation is that this isolate will only sustain an effective plug as long as nutrient medium is continuously supplied to sustain polymer

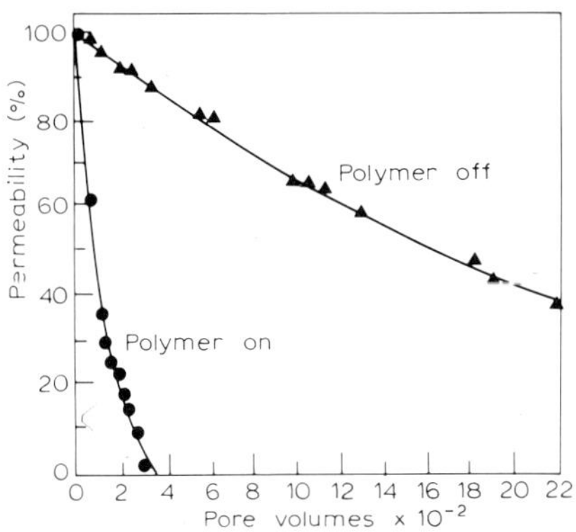

Fig. 7-15. The effect of soluble dextran polymer production on plugging in the model core system by anaerobic wild isolate 18-1. Live culture broth from a medium containing sucrose had polysaccharide formation turned "on" while that from a medium containing glucose/fructose had polysaccharide formation turned "off". Cell densities were about 6–10^8 cells/ml and elapsed times for the runs were 4 and 5$\frac{1}{2}$ hours for polymer "on" and "off", respectively.

formation. When the influent medium is replaced by brine or water, rapid restoration of the permeability is observed presumably due to the wash-out of the soluble polymer along with the bacterial cells. Thus, whereas even a soluble polysaccharide can greatly enhance the rate and extent of microbial plugging, the stability of the plug formed will depend on the nature of the polysaccharide and other exopolymers produced. This observation is reminiscent of the speculations of Mitchell and Nevo (1964), who suggested that the stability of the live bacterial plugging in sand-packed columns was related to the complexity of the polysaccharides formed. Obviously, the immobile glycocalyx of the mature biofilm would represent a recalcitrant microbial plugging problem, while the situation found for the soluble dextran-producing *Leuconostoc* species 18 1 might be more akin to field situations where a soluble biopolymer such as xanthan was being injected for EOR purposes.

IMPLICATIONS

Waterflooding operations

Except where special precautions such as biocide treatment are taken, it is apparent that live bacteria will enter the formation face via the injection well. This is true even for apparently clear waters taken from freshwater wells. It is also apparent that bacteria can plug even very permeable structures having a large average pore throat size.

Two mechanisms of plugging can be distinguished: (1) particulate plugging by the microbial cells themselves, and (2) viable bacterial plugging through biofilm formation. By either mechanism, plugging is concentrated in the formation face immediately adjacent to the wellbore.

In particulate plugging, this is due to the retention of aggregates or clumps of cells and possibly other debris, to which bacteria are attached, in the near-wellbore zone. In the case of biofilms, the continual introduction of oxygen and fresh nutrients in the injection waters favors those bacteria which attach and proliferate in this zone.

The relative importance of these two mechanisms is a function of pore throat size distribution, with tighter formations being more sensitive to particulate plugging. Biocide treatment of the injection waters inhibits the establishment of biofilms in a new injection well, but may not be effective in older wells where biofilm development is already extensive (Ruseska et al., 1982).

MEOR

Based on the above-discussed examples, the following criteria are suggested for viable bacterial cultures being developed as MEOR agents. Successful injection requires the following:

(1) Bacteria must be of a size appropriate to the underground formation; preferably, as small as possible and singly dispersed. The size depends on the pore throat size distribution in the formation face.

(2) Bacteria must not actively produce extracellular polysaccharides or other exopolymers even if these are soluble.

(3) Bacteria should not adhere to the rock surface and form biofilms through glycocalyx production.

One further criterion not discussed here is that the culture must not be capable of generating gas bubbles in the formation face during injection, because these will occupy pore space and in effect reduce the permeability of the formation to the aqueous phase (Jenneman et al., 1982).

Polymer flooding operations

The polymer often used in EOR is xanthan gum which is a bacterial glycocalyx polymer of a chemical composition and chain length that is typical of this class of extracellular molecules (Costerton et al., 1981). The microbial degradation of xanthan gum is now very well documented (Cadmus et al., 1982) and the microbial degradation of polyacrylamides has also been observed (Grula and Sewell, 1983). The initial problem to be faced in a polymer flooding operation is the proliferation of polymer-degrading bacteria in surface handling facilities and the transfer of these potentially troublesome organisms into the injection well. Commercial EOR polymers contain combinations of biocides to prevent bacterial growth, but it must be clearly established that bacterial proliferation does not occur following dilution of the concentrated material with water of doubtful quality.

Most polymer flood EOR operations make use of water injection wells that have been used extensively in secondary waterflood operations. Thus, one must assume that the boreholes and proximal formations of these wells are already heavily fouled by bacteria that have developed very extensive biofilms. Because these pre-existing populations may adsorb the polymer into their biofilms, and many act as a bacterial reservoir to "seed" the initial moving polymer "front" with bacteria, a fastidious clean-up of proposed polymer injection wells would seem to be indicated. The flooding polymer must be protected from microbial degradation, because even minor chain cleavage produces profound losses of viscosity, and biocides proposed for this use must be shown to be effective in the presence of the polymer, of the formation water, and of bacterial populations growing in biofilms in a solid matrix. The fouled core model (Shaw, 1982) would appear to be ideally suited to the purpose of testing biocides for their ability to protect the polymers to be used in EOR, and logic would dictate that biocides should be used in high initial slug doses so that control will be maintained even as adsorption and time-dependent denaturation reduce the effective biocide concentration in the moving polymer front.

Recent unpublished studies have indicated that xanthan gum is indeed degraded to form low-viscosity fragments, within the formation, but that an overall "push" on the oil-bearing stratum has sometimes been achieved. These unconfirmed and unpublished data raise the very interesting possibility that refractory nutrients and bacteria may be carried into the "fingering" zones, responsible for water breakthrough in secondary recovery with subsequent plugging of these water channels through bacterial growth and biofilm development. To initiate EOR using degradable biopolymers without firm evidence of the long-term control of bacterial activity is naive in the extreme, because the basic precepts of microbial ecology dictate that a complex series of population changes will be "set in train" by the introduction of a potential nutrient into the formation. Given the high cost of polymer flooding, careful laboratory study using the polymer, the formation water, and the formation matrix would seem to be required prior to field application.

REFERENCES

Allison, L.E., 1947. Effect of microorganisms on permeability of soil under prolonged submergence. *Soil Sci.*, 63: 439–450.

Anonymous, 1972. Bacteria have destroyed 10% of world's crude. *World Oil*, February, pp. 28–29.

Ashirov, K.B. and Szonova, I.V., 1962. Biogenic sealing of oil deposits in carbonate reservoirs. *Mikrobiologiya*, 31: 680–683.

Beck, J.V., 1947. Prevention of microbiological growth in water flood operations. *Prod. Mon.*, 11: 21–26.

Bitton, G. and Marshall, K.C., 1980. *Adsorption of Microorganisms to Surfaces*. John Wiley and Sons, Toronto, Ont.

Cadmus, M.C., Jackson, L.K., Burton, K.A., Plattner, R.D. and Slodki, M.E., 1982. Biodegradation of xanthan gum by *Bacillus* sp.. *Appl. Environ. Microbiol.*, 44: 5–11.

Cerini, W.F., Battles, W.R. and Jones, P.H., 1946. Some factors influencing the plugging characteristics of an oil-well injection water. *Pet. Technol. Tech. Publ.*, 2028: 1–12.

Costerton, J.W., 1981. Effects of antibiotics on adherent bacteria. *CIBY-GEIGY Symp., Action of Antibiotics in Patients, Florence, July 19*, preprint.

Costerton, J.W. and Geesey, G.G., 1979a. Which populations of aquatic bacteria should we enumerate? In: J.W. Costerton and R.R. Colwell (Editors), *Native Aquatic Bacteria: Enumeration, Activity and Ecology*. American Society for Testing and Materials, Washington, D.C. ASTM STP695, pp. 7–18.

Costerton, J.W. and Geesey, G.G., 1979b. Microbial contamination of surfaces. In: K.L. Mittall (Editor), *Surface Contamination*. Plenum, New York, N.Y., pp. 211–221.

Costerton, J.W., Geesey, G.G. and Cheng, K-J., 1978. How bacteria stick. *Sci. Am.*, 238: 86–95.

Costerton, J.W., Irvin, R.T. and Cheng, K.-J., 1981. The bacterial glycocalyx in nature and disease. *Annu. Rev. Microbiol.*, 35: 299–324.

Davis, J.B., 1967. *Petroleum Microbiology*. Elsevier, Amsterdam.

Geesey, G.G. and Costerton, J.W., 1979. Microbiology of a northern river: bacterial distribution and relationship to suspended sediment and organic carbon. *Can. J. Microbiol.*, 25: 1058–1062.

Geesey, G.G., Mutch, R., Costerton, J.W. and Green, R.B., 1978. Sessile bacteria: an important component of the microbial population in small mountain streams. *Limnol. Oceanogr.*, 23: 1214–1223.

Gruesbeck, C. and Collins, R.E., 1982. Entrainment and deposition of fine particles in porous media. *Soc. Pet. Eng. J.*, 22: 847–856.

Grula, M.M. and Sewell, G.W., 1983. Microbial interactions with polyacrylamide polymers. In: E.C. Donaldson and J.B. Clark (Editors), *Proceedings, 1982 International Conference on Microbial Enhancement of Oil Recovery*. U.S. Department of Energy, NTIS, Springfield, Va.

Gupta, R.P. and Swartzendruber, D., 1962. Flow-associated reduction in hydraulic conductivity of quartz sand. *Soil Sci. Soc. Proc.*, pp. 6–10.

Hart, R.T., Fekete, T. and Flock, D.L., 1960. The plugging effect of bacteria in sandstone systems. *Can. Min. Metall. Bull.*, 53: 495–501.

Hensel, W.M., Jr., Sullivan, R.L. and Stallings, R.H., 1981. Understanding and solving injection well problems. *Pet. Eng. Int.*, May, pp. 155–170.

Jang, L.-K., Findley, J.E. and Yen, T.F., 1983a. Preliminary investigation on the transport problems of microorganisms in porous media. In: J.E. Zajic, D.G. Cooper, T.R. Jack and N. Kosaric (Editors), *Microbial Enhanced Oil Recovery*. PenWell Books, Tulsa, Okla., pp. 45–49.

Jang, L.-K., Sharma, M.M., Findley, J.L., Chang, P.W. and Yen, T.F., 1983b. The transport of bacteria through porous media. In: E.C. Donaldson and J.B. Clark (Editors), *Proceedings, 1982 International Conference on Microbial Enhancement of Oil Recovery*. U.S. Department of Energy, NTIS, Springfield, Va., pp. 60–70.

Jenneman, G.E., Knapp, R.M., McInerney, M.J., Menzie, P.E. and Revus, D.E., 1982. Experimental studies of in situ microbial enhanced oil recovery. *3rd J. Symp. on Enhanced Oil Recovery, SPE-DOE, Tulsa, Okla.*, SPE/DOE 19789.

Kalish, P.J., Stewart, J.A., Rogers, W.F. and Bennett, E.O., 1964. The effect of bacteria on sandstone permeability. *J. Pet. Technol.*, 16: 805–814.

Lada, A., 1959. Microorganisms are bad actors in water flooding. *Oil Gas J.*, 57: 93–95.

McCalla, T.M., 1950. Studies of the effect of microorganisms on rate of percolation of water through soils. *Soil Sci. Soc. Proc.*, pp. 182–186.

McCoy, W.F., Bryers, J.D., Robbins, J. and Costerton, J.W., 1981. Observations of fouling biofilm formation. *Can. J. Microbiol.*, 27: 910–917.

Mitchell, R. and Nevo, Z., 1964. Effect of bacterial polysaccharide accumulation on infiltration of water through sand. *Appl. Microbiol.*, 12: 219–223.

Myers, G.E. and McCready, R.G.L., 1966. Bacteria can penetrate rock. *Can. J. Microbiol.*, 12: 477–484.

Myers, G.E. and Slabyj, B.M., 1962. The microbiological quality of injection water used in Alberta oil-fields. *Proc. Mon.*, May, pp. 12–14.

Raleigh, J.T. and Flock, D.L., 1965. A study of formation plugging with bacteria. *J. Pet. Technol.*, 17: 201–206.

Ruseska, I., Robbins, J., Costerton, J.W. and Lashen, E.S., 1982. Biocide testing against corrosion-causing oil-field bacteria helps control plugging. *Oil Gas J.*, March, pp. 253–264.

Shaw, J.C., 1982. *Bacterial Fouling of a Model Core System*. M.Sc. Thesis, Department of Geology and Geophysics, University of Calgary.

Tosteson, T.R., Revuelta, R., Iman, S.H., Axtmayer, R.W., DeVore, D., Ballantine, D.L., Sasseer, D.S., Morgan, T.O. and Rivera, C., 1982. OTEC biofouling, corrosion and materials study from a moored platform at Punta Tuna, Puerto Rico, II. Microbiofouling. *Ocean Sci. Eng.*, 7: 21–73.

Trautenberg, G.A., 1965. Don't underestimate role of aerobic bacteria. *Oil Gas J.*, January, pp. 85–87.

Ward, D.M. and Brock, T.D., 1978. Anaerobic metabolism of hexadecane in sediments. *Geomicrobiol. J.*, 1: 1–10.

Chapter 8

BACTERIAL MIGRATION THROUGH NUTRIENT-ENRICHED SANDPACK COLUMNS FOR IN-SITU RECOVERY OF OIL

LONG-KUAN JANG, TEH FU YEN, GEORGE V. CHILINGARIAN and ERLE C. DONALDSON

INTRODUCTION

Except for the proposed use of microbial cells and extracellular slimes to selectively plug the highly permeable zones, microbial enhancement of oil recovery (MEOR) methods mainly utilize the metabolites (biosurfactant, biopolymer, organic acid, and biogas) generated in situ or ex situ by bacteria to improve the oil-phase mobility. A well known example of an ex-situ MEOR process is the injection of xanthan gum separated from the growth culture of *Xanthomonas* to thicken the flooding water.

Inasmuch as in-situ MEOR is mainly targeted toward the residual oil left after primary production or secondary production by waterflooding, its success depends strongly on the penetration and the stability of recovering agents as do most tertiary recovery processes. In other words, in order to contact trapped oil with bacteria that have favorable oil displacement properties, the microbes must be transported from a wellbore to locations deep within the reservoir.

Early research indicated that the penetration of the selected bacterial species is rather poor (Hart et al., 1960; Jang et al., 1983b, c) as the bacterial suspension is continuously injected into a clean porous rock. Recent studies, however, showed that the presence of oil in the sandstone core can facilitate bacterial penetration (Jang et al., 1983a). Bacterial spores of *Bacillus subtilis* and *Clostridium acetobutylicum* were found to penetrate more easily than vegetative cells (Jang et al., 1983a, b). Certain chemicals added to the suspending medium can also improve the penetration of bacteria.

In most of the research work, bacterial cells were separated from a growing culture and resuspended in nutrient-free electrolyte solution. The "transport" of cells through the porous media was defined in terms of the extent of penetration as the suspension was continuously injected into the core, essentially because the activity of cells was low and very slow cellular multiplication occurred. A deep-bed filtration model was used to model the transport phenomena in this case (Jang et al., 1983a).

It is generally believed that one advantage of using in-situ MEOR methods is the in-situ generation of metabolites. It is also known that bacteria are able to diffuse through porous rock (Myers and McGready, 1966) or migrate through porous media in the absence of pumping if conditions are favorable for growth. Recently, some progress has been made using a double-flask apparatus to investigate exclusively the rate of migration of *Pseudomonas putida* (ATCC 12633) and *Bacillus subtilis* through stagnant nutrient broth filling a clean Berea Sandstone core (Jang et al., 1983a). It was found that *B. subtilis* migrated faster (1.5 inches/day) than *P. putida* (0.7 inches/day).

The writers investigated the migration and growth of cells through nutrient-enriched, oil-containing reservoir rocks. A process of bio-"huff-and-puff"–nutrient flooding was designed to reveal the effect of in-situ activity and cellular migration on the oil recovery efficiency. An aerobic, biosurfactant-producing Gram-positive rod, *Bacillus* sp., was used by the writers. Inasmuch as metabolites such as biogas and biosurfactant are produced in situ as nutrients and metabolized by cells, a combination of several possible mechanisms such as solution biogas drive, biosurfactant flooding, and alteration of capillary pressure is tentatively proposed for the oil recovery in the so-called bio-"huff-and-puff" process. It also reveals the bacterial transport at various stages of operation, i.e., initial distribution of cells during inoculation, migration during static incubation (cells being pushed toward exit by solution biogas drive) and the growth and migration during makeup nutrient flooding. The details of the process are described in the next section.

MATERIALS AND METHODS

Crude oil

Heavy crude oil from Ranger Zone, Long Beach, California, was used. It has an API gravity of 17°. The viscosity of the crude is 148 cP at 25°C (Chan, 1980).

Microbial species

The culture used by the writers was isolated by Dr. J.E. Findley at the University of Southern California from garden soil. Five grams of a soil sample were cultivated in nutrient broth. The culture was found to consist of a Gram-positive, motile rod as determined by optical microscopy, which formed endospores as the culture aged. The cells were aerobic forming a pellicle on liquid medium. Rough colonies were found to appear on the surface of nutrient agar. The culture was mesophilic and fermented glucose. The above observations are consistent with descriptions of *Bacillus* sp. in Bergey's Manual of Determinative Bacteriology (Buchanan and Gibbons, 1974). When the isolate was cultured in nutrient broth with a magnetic bar stirring the liquid, a thick layer of white foam appeared on the surface of the

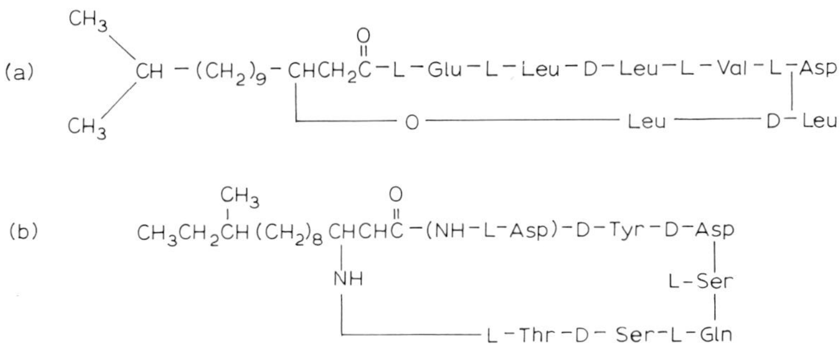

Fig. 8-1. (a) The structure of surfactin or subtilysin, a lipopeptide isolated from *Bacillus subtilis* (after Cooper and Zajic, 1980). (b) One of the surfactin analogs isolated from *Bacillus subtilis* containing an α-amino fatty acid instead of a β-hydroxy fatty acid.

medium during exponential growth. This foam indicated that the culture formed biosurfactant. *Bacillus subtilis* is known to produce biosurfactants, the chemical structure of which is well documented (Cooper and Zajic, 1980) and is shown in Fig. 8-1. No further identification on the biosurfactant produced by the culture was performed, although the crude product was isolated and studied by R. McDavid in 1974 (unpublished). The culture was mainly used to test and demonstrate the process of bio-"huff-and-puff" in the preliminary laboratory study, but will not necessarily be employed in future field tests. The selection of microbial species for actual recovery in candidate petroleum formations remains to be critically evaluated.

Culture medium

The medium was prepared by dissolving 8 g of BBL Nutrient Broth in one liter of distilled water and sterilizing by autoclaving at 125°C for 20 minutes.

Porous media

The oil-containing sandpack column used by the writers to simulate the reservoir rock was prepared by the following procedure. Washed Ottawa sand (composed mainly of silica as determined by EDAX (Energy Dispersive Analysis of X-ray)) was gradually introduced into a Lucite plastic column (28 cm long by 2.54 cm in diameter) with simultaneous manual shaking and tapping for even packing. The average grain size was determined to be 100 mesh by optical microscopy. The sandpack column was evaluated by a vacuum pump for 10 minutes, followed by closing of the valves at both ends of the column. It was later saturated with sterile brine solution containing 1000 mg NaCl/kg (ppm) by aspirating the solution through one end with the other end remaining closed until no more brine entered the column. Twenty pore volumes (PV, 1 PV being the volume of brine that fills the pore space of the column) of 75% alcohol followed by 20 PV of sterile brine were then pumped through the sandpack column to sterilize and flush it. After that, 20 PV of crude oil (unsterilized) was pumped into the column to displace brine until the effluent contained only a very small amount of brine; any water in the porous passages was presumably discontinuous connate water. Twenty pore volumes of sterile brine were again pumped into the column to displace crude oil, until the effluent contained a very small amount of the crude oil. This step was equivalent to secondary waterflooding. The oil remaining in the column after secondary waterflooding (called "initial residual oil" here) was determined to be approximately 20% of 1 PV. (The amounts of oil and water in the sandpack column can be easily determined by measuring the volumes of effluent oil and water in the above steps and by performing a simple mass balance. Oil and water in the effluent readily separate in the measuring cylinder. According to the experience of the writers, however, in order to prevent oil from sticking to the glass wall it is advisable to wet the glass wall of the measuring cylinder by a thin layer of non-ionic detergent.) The initial residual oil was the target for further recovery by microbes and its amount was the basis for calculating the final recovery efficiency. The oil remaining in the column after microbial recovery or nutrient flooding is called "final residual oil" here. The porosity of the clean sandpack column was 39% of the bulk volume of the column. The permeability of the oil-containing column was 4 D. The pumping rate was maintained at 4 PV/day throughout the experiment.

Process description

Fig. 8-2 shows the schematic diagram of the experimental apparatus for testing the process of bio-"huff-and-puff". To supply the nutrient for bacterial growth, about 4 PV of fresh nutrient broth (flask 4 in Fig. 8-2) were injected into the oil-containing sandpack column (column 2) prepared as described in the previous paragraph. About 0.1 PV of *Bacillus* sp. (flask 3) was inoculated through the inlet end (valve *a* and valve *d*) of the column. The inlet end and the outlet end (valve *b*) were then sealed off and the column was incubated for one day at 32°C. (Inoculation plus incubation are the "huff" steps of the process.) Since both ends of column 2 were sealed off during incubation, biogas generated during bacterial growth increased the pressure (approximately 15 psig as indicated by a pressure gauge) inside the column and was readily dissolved in the fermentation broth. The biogas dissolved in the fermentation broth can, therefore, be regarded as "solution gas" in the sandpack column. At the end of the incubation period, the outlet end was opened and the pressure was released accordingly. Upon expansion, solution gas separated from the aqueous phase and served to push a certain amount of fluid containing oil, fermentation broth, and cells outside the column. (The solution-gas drive is the "puff" step of the process.) The oil contained in the produced sample was treated and weighed following the procedure described in the next section.

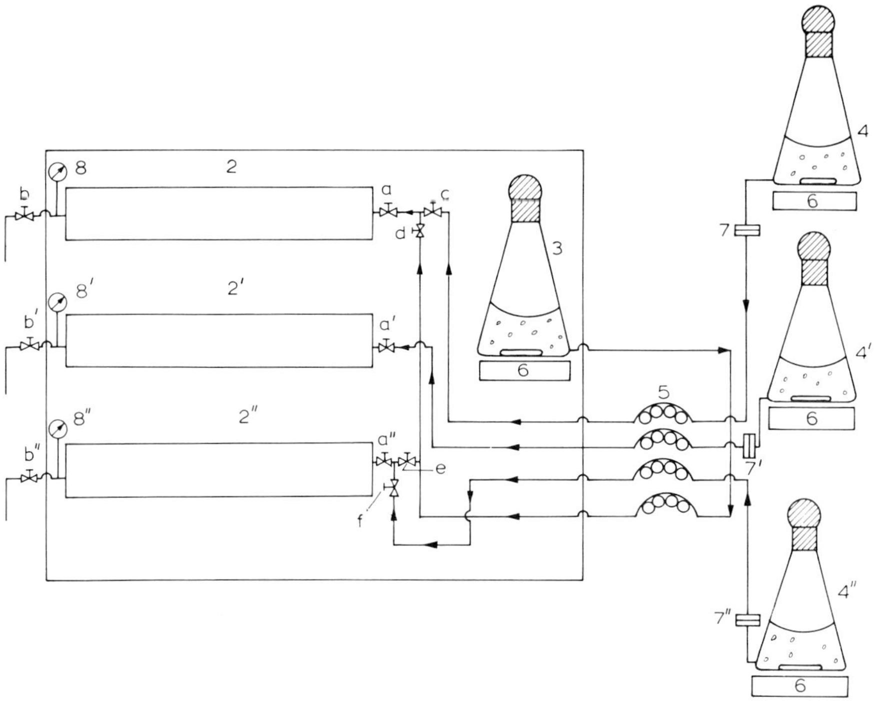

Fig. 8-2. Schematic diagram of the apparatus for recovering residual oil from the sandpack columns. *1* = oven (incubator); *2* = oil-containing sandpack column (for testing recovery by *Bacillus* sp.); *2′* = oil-containing sandpack column (first control run for testing recovery by indigenous bacteria); *2″* = clean sandpack column (second control run for monitoring the growth and migration of *Bacillus* sp.); *3* = *Bacillus* sp. culture; *4*, *4′*, *4″* = fresh nutrient broth for columns 2, 2′, and 2″, respectively; *5* = 4-channel tubing pump; *6* = magnetic stirrer; *7* = membrane filter; *8*, *8′*, *8″* = pressure gauges; *a–f*: stainless steel valves.

To determine whether the fermentation broth in the column contained residual nutrient that could support further growth of bacteria after the first "huff-and-puff" cycle, additional cycles of sealing, incubation, and pressure release were applied until the magnitude of production in the "puff" step became nil. The decrease in the rate (strength) of production was presumably due to the depletion of nutrient and/or production of inhibitory products, and the aging of cells in the column. To stimulate the activity of the cells, about 4 PV of fresh nutrient (flask 4) were fed into the column (valves *a*, *b*, and *c* were open while valve *d* was closed in this step). During this step of continuous make-up nutrient flooding, the oil displaced by the flooding medium was also collected and treated. Following the injection of makeup nutrient, the column was again incubated for one day with both ends closed, followed by release of pressure at the outlet end and the collection of produced fluid. Several additional cycles of makeup nutrient flooding, incubation, and pres-

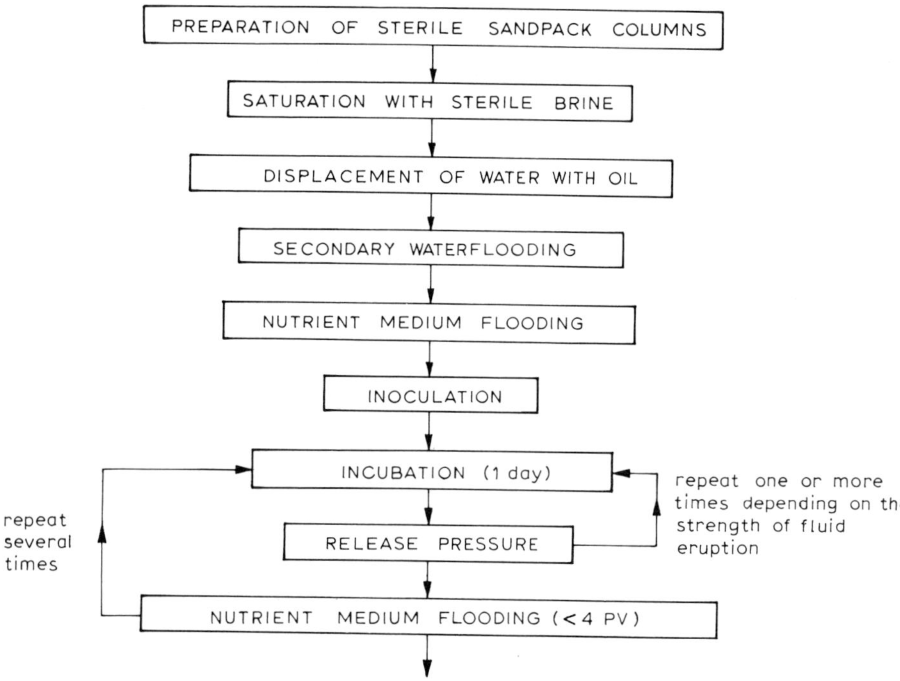

Fig. 8-3. Flow diagram of the bio-"huff-and-puff" process for recovering residual oil from sandpack columns using *Bacillus* sp. In the first control run, the step of inoculation was omitted. In the second control run, the column was not charged with oil.

sure release were applied to obtain maximum recovery of oil from the column. The flow diagram for the bio-"huff-and-puff" recovery process and its two control runs are shown in Fig. 8-3.

The first control run (column 2′ in Fig. 8-2) was operated in a manner similar to the above-mentioned recovery experiment except that the column was not inoculated. The purpose was to determine whether or not bacteria indigenous to the crude oil could grow in the nutrient medium.

In the second control run, 0.1 PV of culture was introduced (inoculated) into a clean column (column 2″ in Fig. 8-2), which was not charged with oil (i.e., the steps of injecting oil and displacing oil with sterile brine in Fig. 8-3 were omitted). The main purpose of this control run was to monitor the effluent bacterial count in order to obtain information regarding the migration and the growth of *Bacillus* sp. in the clean sand pack. Previously, the writers have shown that by charging a Berea Sandstone core with heavy crude oil from Long Beach, California, the penetration of bacteria could be facilitated as the nutrient-free suspension was continuously pumped in (Jang et al., 1983a). Similarly, the results obtained from the second control run, with no oil present in the column, provided a baseline for the growth and migration of cells through static nutrient in the porous media. Also, comparing the bacterial activity in column 2 and column 2″, as indicated by the strength of solution-gas drive, it was possible to judge if the presence of oil and/or bacteria indigenous to the oil can influence the activity of the inoculated bacteria in the oil-containing sandpack column. The bacterial count in the column effluent was determined by plating on nutrient agar.

The writers investigated several oil recovery methods, including caustic flooding and microbial enhanced oil recovery processes. All experiments were performed at least twice. The standard procedures of preparing oil-bearing porous media, recovering residual oil, and quantifying recovery efficiency were followed.

Treatment of oil samples

Since the oil used in the experiment was a heavy crude oil which tended to adhere to the glass wall of beakers and measuring cylinders, it was fairly difficult to accurately measure the amount of oil recovered in the produced fluid volumetrically. To solve the problem, toluene was added to each sample. The oil contained in each sampling beaker was readily dissolved in toluene and separated from the aqueous medium by brief shaking and stirring. The hydrocarbon layer (i.e., toluene plus crude oil) was then easily removed with a pipette. To evaporate the toluene from each sample, the sample was placed in a fume hood for about one day. The sample was further dried in vacuo at 10 psi and at 40°C overnight to remove residual toluene and moisture. The net weight of the "dry oil" sample was determined by weighing on an analytical balance.

The final residual oil remaining in the sandpack column after a recovery experiment was determined similarly. At the end of the experiment, the caps on both ends of column 2 and column 2′ were opened. The final content of each column was removed and collected in a large beaker. The residual oil was separated from the sand grains by extracting with toluene until the sand grains were clean. To evaporate the large amount of toluene, the combined extract was first heated under vacuum in a rotary evaporator. The bottom of the rotary evaporator was immersed in a water bath heated to approximately 55°C. The evaporator was evacuated by using a water aspirator. The concentrated extract was then dried and quantified following the procedure described above.

Calculation of recovery efficiency

The percentage of oil recovered was calculated as the ratio of the total dry weight of oil recovered (A) to the dry weight of the initial residual oil after secondary brine flooding ($A + B$, where B is the dry weight of the final residual oil in column 2 or column 2′).

The course of cumulative oil recovery (dry weight) from column 2 is shown in Fig. 8-4. Since both ends were sealed off during incubation and there was no discharge of oil, the incubation periods can, therefore, be represented by horizontal sections on the curve in Fig. 8-4 (for instance, the first 24-hour incubation period is represented by the section *RT*). The amount of oil pushed out by the pressure of solution gas, each time the exit valve was opened after incubation, is represented by the height of the dashed arrow pointing upwards (for instance, the amount of oil pushed out at the 48th hour is approximately 0.8 g, represented by the height of *TS*). The difference between points *P* and *Q* in Fig. 8-4 is the total dry weight of oil recovered in the period between the 168th hour and the 192nd hour, during which the column was continuously flooded with fresh, aerated nutrient broth. Assuming that the production rate in this period is constant, the cumulative recovery for this continuous flooding period is then approximated by a straight, upward-sloping line. Likewise, the cumulative recovery history for other nutrient-flooding periods can be approximated by straight, upward-sloping lines. Of course, more accurate oil recovery history for the nutrient-flooding periods would be obtained if one subdivides each nutrient-flooding period, collects samples for the sub-periods, and determines their cumulative weight.

In Fig. 8-4, the beginning of nutrient injection following the completion of secondary waterflooding is chosen as time zero. Approximately 1.2 g (dry weight, the difference between points *Q* and *R* in Fig. 8-4) of oil was displaced by nutrient medium prior to inoculation of bacteria (point *R*). The oil displaced during this

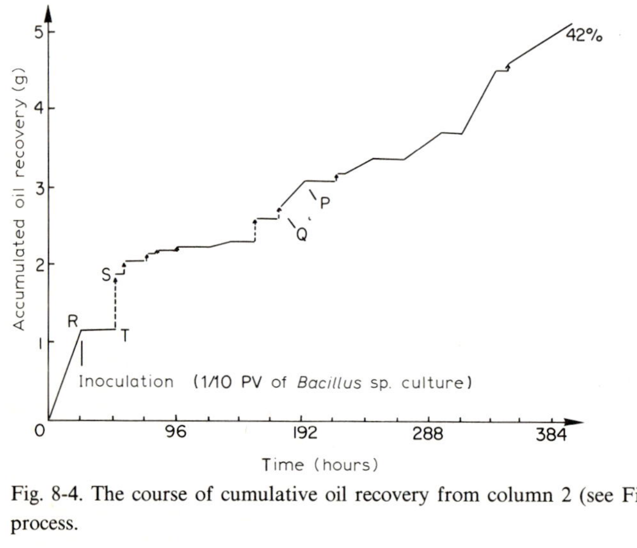

Fig. 8-4. The course of cumulative oil recovery from column 2 (see Fig. 8-2) in the bio-"huff-and-puff" process.

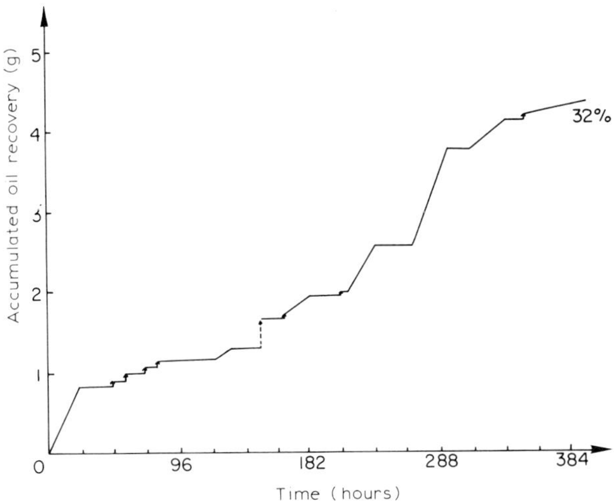

Fig. 8-5. The course of cumulative oil recovery in the first control run (column 2′ in Fig. 8-2) of the bio-"huff-and-puff" process.

period may be due to the detergency of some of the compounds in the nutrient broth. The oil recovered from column 2 at point T was about 0.8 g (dry weight, the difference between points T and S), whereas in the first control run (column 2′) a much smaller amount was obtained (see Fig. 8-5). Also, it was observed that the strength of solution-gas drive in column 2 was much higher than that in column 2′ (i.e., the first control run). About 5 ml of fluid was pushed out from the former, whereas only a very small amount (less than 1 ml) was pushed out from the latter. It was apparent that *Bacillus* sp. in column 2 was more active than the indigenous bacteria in column 2′, especially in the early stage of the experiment. The oil production in the first control run (column 2′, Fig. 8-5) gradually increased in the latter half of the experiment, probably resulting from the growth of indigenous bacteria. The average recovery efficiencies from column 2 and column 2′ were 42% (or 5.04 g based on the 12.1 g of initial residual oil remaining in the sandpack column after secondary waterflooding) and 32% (or 3.9 g), respectively.

In the second control run, bacterial concentrations in the effluent samples collected during nutrient flooding or when the exit valve (valve b'') was opened after incubation were around 10^6/ml throughout the experiments as determined by plating on nutrient agar. It is worth noting that the bacterial count in the first produced sample (48th hour), when the outlet end (valve b'') was opened, reached 2×10^5/ml only one day after inoculation through the inlet end. Also, the strength of solution-gas drive was about the same as that in column 2. The highest pressure developed during incubation of column 2″ was approximately 15 psig, and the volume of the first produced sample was about 5 ml. Inasmuch as the column was initially sterile, it is apparent that *Bacillus* sp. cells were able to grow and migrate through the sandpack column saturated with nutrient medium. Also it is known that *Bacillus* sp. adsorbs strongly onto rock surfaces and a large contact area exists between the aqueous phase and the porous media. It can be anticipated that cells collected in the effluent samples only represent a small fraction of the total population.

No bacterial enumeration of the effluent samples from column 2 and column 2′ was conducted because the samples were used totally for quantifying the amount of oil produced. Although the results from column 2″ do not necessarily apply to column 2, the evidence for vigorous growth and migration of cells in column 2 was clearly demonstrated by comparing the intensity of biogas production and the amount of fluid pushed out. Had vigorous growth and bacterial migration through column 2 not occurred, only a very small amount of biogas would have been produced. The evidence for growth and migration of *Bacillus* sp. through column 2 was further supported by comparing the results of column 2 and column 2″. Apparently, the indigenous bacteria assumed to have been present did not hinder the growth and migration of inoculated *Bacillus* sp. Otherwise, one would not have observed a strong biogas production and a high oil recovery efficiency in column 2. In short, the presence of oil and indigenous bacteria does not hinder the growth and migration of inoculated bacteria.

DISCUSSION

The process of repeated cycles of bio-"huff-and-puff" developed by the writers shows several advantages. First, although only a small amount (0.1 PV) of bacterial culture is inoculated and the initial distribution of cells is limited within the inlet region, the further penetration of cells into the column was achieved by growth and migration during static incubation. It is likely that the penetration of *Bacillus* sp. is enhanced by the motility of cells which possess peritrichous flagella. Second, during static incubation, while cells migrate through the column, nutrients are simultaneously converted into metabolites which can help in mobilizing oil. This process, therefore, eliminates part of the need for continuous pumping of recovery agents in conventional enhanced oil recovery processes. Third, by supplying fresh nutrient to revive the in-situ activity, several cycles of "huff-and-puff" can be applied and, more importantly, several different mechanisms resulting from the bacterial growth and the generation of metabolites can be utilized.

Although the detailed mechanisms for this bio-"huff-and-puff" process are not fully identified, the oil recovery process can be described as follows, using the appropriate traits of the bacterial strains and features of well established results in the field of enhanced oil recovery. As both ends of the sandpack columns were sealed during incubation after inoculation, the biogas generated in this period increased the pressure of the system and is believed to be readily dissolved in water. Once the pressure was suddenly released by opening the exit valve, the dissolved gas expanded, leaving the aqueous phase, and pushed out water and some residual oil.

Cells were also apparently flushed toward the exit, which would definitely aid the penetration of cells in the column. The sand used was composed mainly of silica and, moreover, *Bacillus* sp. are not acid-producing strains. Apparently, no chemical reaction occurred between cellular metabolites and sand grains. The gas-producing ability of *Bacillus* sp. is small and only a slight increase of 15 psig in pressure was observed during incubation. The volumes of the fluid produced at the first pressure release were only about 5 ml, or approximately one-tenth of the pore volume.

B. subtilis is known to be a biosurfactant producer (Zajic and Seffens, 1984). Research has shown that adding the biosurfactants (called surfactin, see Fig. 8-1 for structure) isolated from a growing culture of *B. subtilis* to the flooding water can enhance the efficiency of oil recovery (Zajic and Seffens, 1984). In such an ex-situ process, however, it is necessary to pump flooding water continuously into the reservoir. In the bio-"huff-and-puff" process designed in this study, biosurfactants, if produced, are generated in situ along with biogas. (Although the production of biosurfactants was not directly determined, the white foam appearing on the surface of liquid medium in flask 3 of Fig. 8-2 indicated the presence of biosurfactants.) It is likely that during the first incubation period, the biosurfactants produced in situ altered the interfacial tension between oil and water and wetted the sand grains. This would reduce the presence required to push the oil through the capillary passages in the porous medium. Thus, an appreciable amount of oil was pushed out of column 2 as the exit was opened after the first incubation period, although the gas-producing ability of *Bacillus* sp. is small compared with other more vigorous gas-producing species.

The makeup nutrient flooding served to displace the oil which had previously been released by biosurfactants during static incubation and pushed out from dead ends of the porous media by solution-biogas drive. Additionally, nutrient flooding displaced metabolic wastes in the column and stimulated the activity of the cells. The cells, therefore, could grow and migrate further into the porous media while more metabolites were generated in situ. Thus, several cycles of "huff-and-puff" followed by nutrient flooding can be applied to recover more residual oil. It is apparent from Fig. 8-4 that in the latter half of the experiment the amount of oil produced by solution-gas drive diminished, while nutrient flooding (likely coupled with cellular multiplication and generation of biosurfactants) played a major role in the recovery of oil.

The composition of produced biogas remains to be studied. There is no doubt that the biogas produced in column 2″ was due totally to the metabolism by the inoculated bacteria. As mentioned earlier, no gas can result from chemical reactions between sand grains and bacterial metabolites. Since nutrient broth was used, the culture is likely to produce ammonia from proteinaceous components in the broth along with respiratory carbon dioxide. If enough ammonia is produced during the first incubation period in a closed system such as column 2″, the pH of the fermentation broth in the column is likely to be raised to 8 or 9 resulting from the dissolution of ammonia. If this is the case, most or all of the respiratory CO_2 produced will become dissolved in the broth as HCO_3^- and CO_3^{2-}. The pH of the first produced sample from column 2″ was determined to be 6.8. However, it does not represent the true condition in the column. Prior to opening the exit valve (valve *b*″) at the 48th hour, the fluid in the column was under an elevated pressure of 15 psig, whereas the measurement of pH of the produced sample was made at normal atmospheric pressure. The concentration of dissolved gas (e.g., ammonia and CO_2) was different and, therefore, the pH can be different in the two cases. Further detailed investigation into the metabolism of the strains used for MEOR is needed so that the recovery mechanisms in MEOR processes can be better understood.

The term "solution-gas drive" is frequently used in petroleum production. In some reservoirs, gases such as methane, ethane, and carbon dioxide are dissolved in petroleum under elevated pore pressures. The pressure decreases as the fluid flows toward the production well or after the primary production has proceeded for certain periods of time. The gases originally dissolved in the petroleum will separate from the oil phase, expand, and aid in pushing oil toward the production well. The biogas (presumably composed mainly of carbon dioxide and ammonia and, perhaps, some methane as discussed in the next paragraph) produced by bacterial cells in the sandpack columns used here is, however, more likely dissolved in the aqueous fermentation broth instead of the oil. This is because only a limited level of compression (ca. 15 psig) results from the generation of biogas in a closed system and the thermodynamic phase behavior at this pressure level favors the dissolution of biogas in the aqueous solution. Actually, during the first incubation period (i.e., between the 24th and 48th hour), no gas-solution separation in the sandpack columns 2, 2′, and 2″ was observed (the lucite plastic column is transparent and the surface of the packing is visible.) As the exit valves *b*, *b*′, and *b*″ were opened after the first incubation period, pore spaces occupied by biogas appeared and quickly spread throughout the column as the fluids in the columns were simultaneously pushed out (the intensity was higher in the case of columns 2 and 2″ and lower with column 2′.) The writers, therefore, use the term "solution-biogas drive" to describe the mechanism of fluid movement and oil recovery as the pressure of the columns is released, although this term is applied slightly differently in the primary recovery of petroleum production.

Judging from the oil recovery efficiency, the amount of produced fluid, and the bacterial count in the effluent, the bacteria indigenous to the Long Beach crude oil did not seem to hinder the activity of *Bacillus* sp. in the bench-scale experiment of bio-"huff-and-puff". It is well known that petroleum formations are easily contaminated with bacteria (i.e., adventitious bacteria) present in drilling fluids and in injection waters (Davis, 1967). Many aerobic and anaerobic bacterial species indigenous to petroleum formations have also been identified (Bushnell and Haas, 1940; Tornabene, 1975; Grula and Grula, 1979). A more careful study is necessary to ensure that the selected species (or a group of several species) will be dominant under the nutrient and reservoir conditions encountered. If one is dealing with a dynamic environment, such as a petroleum formation involving several microbial species, one should envision the formation as an ecosystem. As an example, one can consider the bench-scale system of column 2 and column 2′. Both aerobic and anaerobic indigenous bacteria were present in this experiment. Cells of the inoculated bacterial culture grew and migrated faster through the column than indigenous bacteria, because the former were inoculated at the exponential growth phase. As the inoculated cells multiplied, oxygen may have been depleted in the column, leaving an anaerobic zone behind the migration front of the inoculated culture. The anaerobic cells of the indigenous bacteria may start to grow and metabolize the remaining nutrient and/or the metabolites produced by the inoculated bacteria. Metabolites resulting from anaerobic fermentation, such as methane, organic solvents, or organic acids could then be generated. Moreover, as the carbon source supplied in the nutrient is depleted, cells may further metabolize hydrocarbons in the crude oil. In short, the interaction among inoculated bacteria, indigenous bacteria, adventitious bacteria, nutrients supplied, oil, and porous formation is very complex. The writers designed the experiment mainly to demonstrate the bench-scale test of using a bacterial culture to recover oil in an engineering approach. Some advantages of the bio-"huff-and-puff" process over other conventional tertiary recovery processes were demonstrated. Nevertheless, much work remains to be done on the selection or the manipulation of bacterial species for oil recovery, the identification of both indigenous and adventitious bacteria, and the identification of metabolic products of cells in a complex ecosystem such as a petroleum formation.

Considering the anaerobic conditions in a reservoir and the technical difficulty of supplying oxygen, it could be more feasible to inject anaerobic species such as *Clostridium acetobutylicum* (an acid–gas–solvent producer, which can produce 220 mmol CO_2, 165.9 mmol H_2, 24.8 mmol acetate, 7.1 mmol butyrate, 4.9 mmol ethanol, 47.4 mmol butanol, 22.3 mmol acetone, and 5.7 mmol acetonin per 100 mmol of glucose consumed (Moat, 1979)).

The nutrient medium used here was a high-priced, rich proteinaceous nutrient broth commonly used by microbiologists to cultivate a wide spectrum of aerobic bacteria. In practice, one may want to choose an inexpensive, less concentrated nutrient source such as dilute molasses solution. The growth rate, metabolic products and the recovery mechanisms may be different from those presented here. Some of the methods and the sequences of flooding developed by the writers, however, may be applied.

ACKNOWLEDGMENTS

The research work was supported by the U.S. Department of Energy under contract DE-AS19-81-BC10508. The writers are grateful to Dr. John E. Findley of University of Southern California for supplying the bacterial culture.

REFERENCES

Buchanan, R.E. and Gibbons, N.E. (Co-Editors), 1974. *Bergey's Manual of Determinative Bacteriology.* Williams and Wilkens, Baltimore, Md., 8th ed.

Bushnell, L.D. and Haas, H.F., 1940. The utilization of certain hydrocarbons by microorganisms. *J. Bacteriol.*, 41: 653–673; also *Dep. Bacteriol., Kansas Agric. Exp. Stn. Publ.*, 199: 653–673.

Chan, M., 1980. *Interfacial Activity in Alkaline Flooding Enhanced Oil Recovery*. Ph.D. Dissertation, Department of Chemical Engineering, University of Southern California, Los Angeles, Calif., 75 pp.

Cooper, D.G. and Zajic, J.E., 1980. In: D. Perlman (Editor), *Advances in Applied Microbiology, 26.* Academic Press, New York, N.Y., pp. 339–340.

Davis, J.B., 1967. *Petroleum Microbiology*. Elsevier, Amsterdam, pp. 449–453, 496–498.

Grula, M.M. and Grula, E.A., 1979. Biodegradability of compounds used in enhanced oil recovery. In: *5th DOE Symposium on Enhanced Oil and Gas Recovery and Improved Drilling Technology, 3. Gas and Drilling*. The Petroleum Publishing Company, Washington, D.C., pp. E-2/1–E-2/7.

Hart, R.T., Fekete, T. and Flock, D.L., 1960. The plugging effect of bacteria in sandstone system. *Can. Min. Metall. Bull.*, 53: 495–501.

Jang, L.K., Chang, P.W., Findley, J.E. and Yen, T.F., 1983a. Selection of bacteria with favorable transport properties through porous media for the application of microbial enhanced oil recovery. *Appl. Environ. Microbiol.*, 46(5): 1066–1072.

Jang, L.K., Findley, J.E. and Yen, T.F., 1983b. Preliminary investigation of the transport problems of microorganisms in porous media. In: J.E. Zajic, D.G. Cooper, T.R. Jack and N. Kosaric (Editors), *Microbial Enhanced Oil Recovery*. PennWell Books, Tulsa, Okla., pp. 45–49.

Jang, L.K., Sharma, M.M., Findley, J.E., Chang, P.W. and Yen, T.F., 1983c. An investigation of the transport of bacteria through porous media. In: E.C. Donaldson and J.B. Clark (Editors), *Proceedings, 1982 International Conference on Microbial Enhancement of Oil Recovery*. U.S. Department of Energy, NTIS, Springfield, Va., pp. 60–70.

Karaskiewicz, J., 1968. Recovery of crude oil from reservoirs by the use of bacteria. *Nafta (Katowice, Pol.)*, 24(7): 198.

Moat, A.G., 1979. *Microbial Physiology*. John Wiley and Sons, New York, N.Y., 1st ed., p. 145.

Myers, G.E. and McGready, R.G.L., 1966. Bacteria can penetrate rock. *Can. J. Microbiol.*, 12: 477–484.

Tornabene, T.G., 1976. *Microbial Formation of Hydrocarbons in Microbial Energy Conversion*. Pergamon Press, Oxford, pp. 281–299.

Zajic, J.E. and Seffens, W., 1984. Biosurfactants. *Crit. Rev. Biotechnol.*, 1(2): 87–107.

Chapter 9

OIL DISPLACEMENT IN THE HELE-SHAW MODEL USING MICROBES AND SYNTHETIC SURFACTANTS

J.E. ZAJIC *, T. BAN, A. GURROLA and W. SEFFENS

INTRODUCTION

Initial recovery of oil from a porous medium such as an oil reservoir is typically accomplished by collecting oil which is either flowing because of natural reservoir pressure or by using pumping devices. This is referred to as primary oil recovery. With time and as oil and gas are removed, the natural reservoir pressure decreases. Vast amounts of yet unrecovered oil (often greater than 65%) remain in the reservoir and the question of how to recover such oil inexpensively still remains a technical and economic challenge.

Secondary oil recovery procedures involve water injection to displace oil toward collecting sites. Enhanced oil recovery includes several important physical and chemical processes. The use of microbes and/or their products, with or without synthetic products, is referred to as microbial enhanced oil recovery (MEOR). Many potential processes for MEOR exist, e.g., biosurfactants, biopolymers and biocosurfactants.

Various laboratory models for evaluating oil removal from porous media have been developed. This study employed a Hele-Shaw model to test solutions of commercial surfactants and surfactants produced by bacterial cultures for oil recovery efficiency. The Hele-Shaw model does not purport to be a method of simulating conditions present in a reservoir. It represents a large, simple, single-pore method of evaluating and comparing solutions for the purpose of identifying the important physical factors of oil for displacement from porous media. The types of porous media which have been used for evaluating oil separation have been summarized in detail by Taber (1981).

MATERIALS USED IN THE EXPERIMENTS

Oil

A non-detergent motor oil was used in the experiments. Crude oil also has been investigated but is not reported herein. The viscosity of this oil was 832 cP at 20°C

* Deceased.

and the interfacial tension (γ_i) of the oil against distilled water was 27 mN/m. Operational flow through the Hele-Shaw cell was laminar (Reynold's numbers < 300).

Surfactants

Pluronic surfactants of varying hydrophilic–lipophilic balance (HLB) were used to reduce the interfacial tension between oil and the aqueous phase. These non-ionic surfactants are a series of straight-chain, primary aliphatic oxyalkylated alcohols.

Surfactant solutions tested in the Hele-Shaw model were prepared by adding varying concentrations of the surfactant to either water or a polymer solution of xanthan gum polymer (xanflood, XFL-50712K).

Cultures

Most of the biostudies utilized a high-surfactant-producing culture of *Corynebacterium fascians*. Surfactant production and the surface activity of *C. fascians* may benefit enhanced oil recovery techniques and bitumen separation from tar sands.

C. fascians was grown in a mineral salt–hydrocarbon medium consisting of: 0.4% NH_4NO_3, 0.4% KH_2PO_4, 0.6% Na_2HPO_4, 0.02% $MgSO_4.7H_2O$, 0.0001% $CaCl_2.2H_2O$, 0.0001% $FeSO_4.7H_2O$, 0.000014% EDTA (Na), 0.1% yeast extract, and 6% kerosene. Undiluted as well as diluted cultures of *C. fascians* were tested in the Hele-Shaw test chamber.

APPARATUS

The Hele-Shaw apparatus illustrated in Fig. 9-1 consists of a vessel (two square plates with a thin spacer around the periphery) with diagonal inlet and outlet. The test chamber was filled with fluid through rubber and glass tubing. Displacement

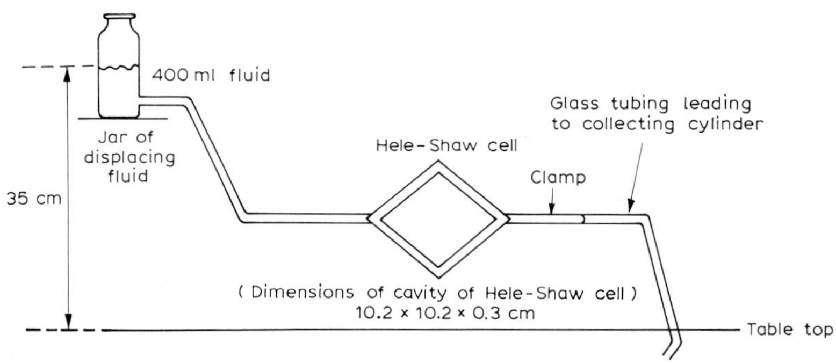

Fig. 9-1. Schematic diagram of the apparatus used (Hele-Shaw model).

fluids were collected in graduated cylinders. A platform was used for adjusting the height of the displacement fluid to achieve a constant pressure head. A spigotted displacement vessel containing oil was used to fill the Hele-Shaw cell (not shown). A magnetic stirrer was used to mix and maintain a uniform concentration of displacement fluid.

The Hele-Shaw cell consists of two closely-spaced glass plates (3-mm spacing) with provisions for displacement of fluid from the single pore. Spacing of the glass plates was accomplished by placing thin glass strips at edges of the plates and gluing them together with a sealant. Stainless steel tubes were placed at opposite corners of the model for inlet and outlet flow. The internal dimensions of the cell chamber was $10.2 \times 10.2 \times 0.3$ cm.

EXPERIMENTAL PROCEDURE

A spigotted reservoir jar is filled with 400 ml of test fluid and raised to a fluid height of 35 cm (hydrostatic head pressure). All displacement fluids were "water-dominated" systems to which surfactants were added as shown in the test experiments. Care must be exercised to remove all air bubbles present in the tubing as such bubbles interfere with or block liquid flow. The Hele-Shaw cell was filled with non-detergent oil from the oil reservoir vessel followed by injection of the test fluids. The oil–water system (the displacing fluid is referred to as water) is held in place with a valve as shown in Fig. 9-1.

To initiate a test, the valve is opened, flow initiated, and a timer started. As would be expected, the initial liquid displaced is totally oil. After the point of water breakthrough, the oil/water ratio becomes very small and decreases with time. With the test system described, 6 graduated cylinders (25 ml each) are filled (150 ml volume) with displacement fluid. Breakthrough usually is observed in the first 25 ml volume. The volumes of oil and water remaining in the glass tubes were collected for complete material balance calculations. The rate of fluid flowing through the system is determined by collecting the effluent fluids and recording the time of collection. Test runs were made in duplicate, and good reproducibility was observed.

The Hele-Shaw model has been used to measure displacement efficiency of an aqueous phase pushing an oil phase within a defined pore geometry (Gogarty and Fosch, 1968). It is used typically to evaluate polymers. Surfactant, polymer–surfactant, and bacterial solutions have been tested herein for evaluating oil displacement ability.

Characterization of the defined geometry of the Hele-Shaw model has been performed in terms of permeability and Reynold's number calculations (Zajic et al., 1983). The permeability to water calculated for the Hele-Shaw model is comparable to permeabilities of soils and loose bed sands (Scheidegger, 1960). Calculated Reynold's numbers were below 300, thus indicating laminar flow.

The Hele-Shaw model has previously been used for the purpose of investigating dynamics and theory of single- and two-phase flow systems (gas–liquid, liquid–liquid) (Hele-Shaw, 1897, 1898; Taber, 1981). The correlation group known as the capillary number may be used for evaluating the oil recovery from the Hele-Shaw model:

$$N_c = v\mu/\gamma_i \tag{9-1}$$

where v is the velocity of displacement by wetting phase, μ is viscosity, and γ_i is the interfacial tension between wetting phase and oil (Saffman and Taylor, 1958). The relationship between the correlation coefficient and the reduction of interfacial tension with its effect on oil displacement in the Hele-Shaw model were studied here. Mobility control evaluations were limited, because viscosity (the predominant factor in mobility control) of displacing fluids in most of these studies was held constant. In this study, the model was used to evaluate the term of the functional group relating to interfacial tension.

EFFECT OF PLURONIC SURFACTANT SOLUTIONS

Under reservoir conditions, surfactants serve to reduce capillary pressure which aids in removing oil from the capillary pores in the porous media. Evaluation of oil displacement ability by surfactant solutions in the single-pore Hele-Shaw model was performed.

Pluronic surfactants (L61, L63, F87, L62 and P75) were used to reduce the interfacial tension between oil and water. Fig. 9-2 shows the oil recovery using different pluronic surfactants added to water. By varying the concentration of surfactant, solutions of varying interfacial tension were produced for each surfactant. The viscosity of the displacement fluid was that of water. For each surfactant solution shown, there is an observable peak in oil displacement at a concentration of 0.002% surfactant. The amount of oil recovered at this peak was 49–50% of the original oil-in-place. The interfacial tensions produced at a concentration of 0.002% are shown in Fig. 9-2. Interfacial tensions at this concentration for most of the surfactants were approximately 15 ± 10 mN/m. Concentrations greater than 0.002% cause lower interfacial tensions, whereas concentrations less than 0.002% cause higher interfacial tensions. Fig. 9-3 compares oil recovery of the pluronic surfactants at a constant concentration of 0.002%. As the observed interfacial tension deviates to either side of the apparent optimal interfacial tension (15 mN/m), oil recovery decreases. The maximum volume of oil displaced was 50%. The farther from this optimal interfacial tension, the less the oil recovery. Although a model of this type cannot represent the conditions existing in a consolidated multiporous medium typical of a reservoir, it may help identify the controlling parameters. The surfactants chosen possessed a wide array of HLB values (a reference index number indicating the hydrophile–lipophile balance) and molecular weights (Adamson, 1976).

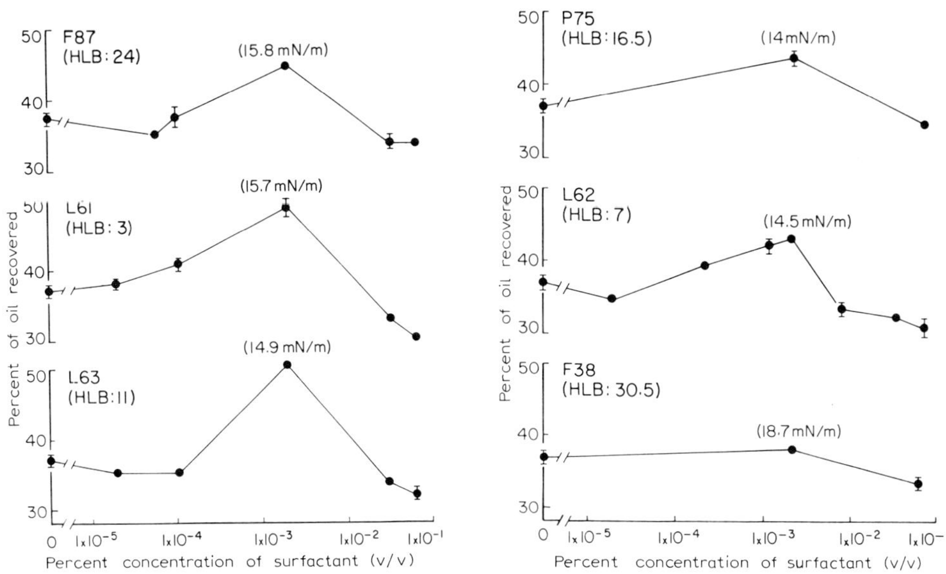

Fig. 9-2. Recovery of oil in the Hele-Shaw model using various surfactant solutions at varying concentrations.

Fig. 9-4 demonstrates the effects of hydrophobicity and molecular weight of surfactants on oil displacement from the Hele-Shaw model. Data points are connected by solid lines. Extreme hydrophobicity and hydrophilicity values of 1.0 and 29, respectively, are not favored for oil displacement in his model. Surfactants L61 (HLB 7) and L63 (HLB 11) gave the highest oil displacement. If the observable trends are HLB dependent, a blend of two surfactants giving an intermediate

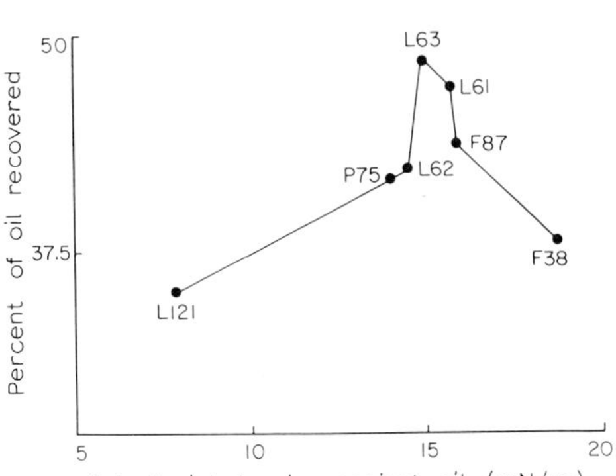

Fig. 9-3. Effect of interfacial tension on oil recovery in the Hele-Shaw model. Concentration of each surfactant solution: 0.002%.

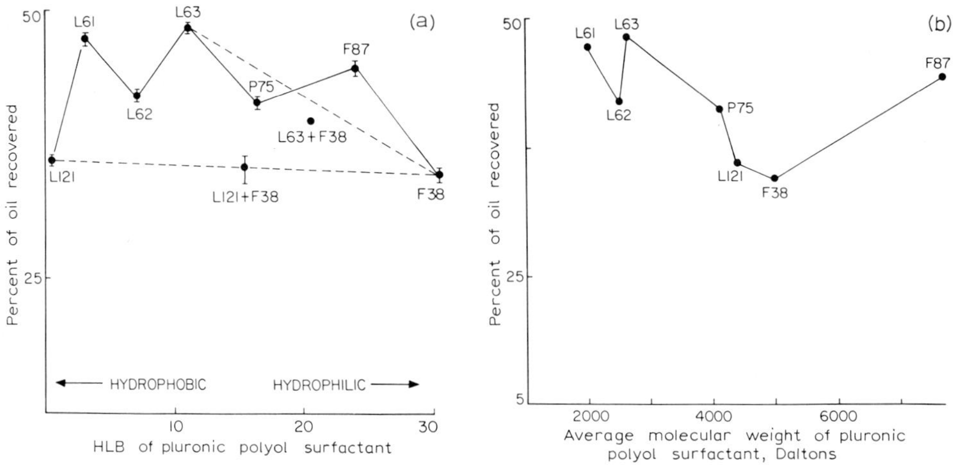

Fig. 9-4. (a) Effect of HLB on oil recovery in the Hele-Shaw model. All surfactant solutions have the same interfacial tension. (b) Effect of molecular weight on oil recovery in the Hele-Shaw model. All surfactant solutions have the same interfacial tension (15 mN/m).

weighted (average) HLB should give the level of oil recovery as a single surfactant of identical HLB. A test of this type was completed using an equal-weight mixture of L121 (HLB 0.5) and F38 (HLB 30.5). This solution should display emulsifying properties similar to a surfactant with HLB of 15.5. The recovery data for this test mixture did not correspond to the data for pluronic P75 surfactant (HLB 16.5). The data point for the mixture lies on the dotted line between the solution components. This probably indicates that oil recovery in the single-pore model is a function of another property, the contribution of which is far greater than HLB. To check these observations, the experiment was repeated with another surfactant solution mixture of F38 (HLB 30.5) and L63 (HLB 11) to determine if the oil recovery is dependent on the average composition of the two surfactants. If no chemical interaction occurs between the surfactants in solution, then an equal mixture of the relatively "good" surfactant L63 with the "poor" one (F38) should give intermediate oil recovery in the test system. This indeed appears to occur although the data point falls somewhat below the dotted line connecting L63 to F38.

Experiments thus far indicate that the HLB and molecular weight of the surfactant are of lesser importance and there appears to be an optimal interfacial tension for oil displacement from the single pore with aqueous solutions at $\gamma_i = 15$ mN/m. This is little surprising in light of the current research activity seeking ultra-low interfacial tensions for enhanced oil recovery (also see Taber, 1969). It probably is a characteristic of the model used. As already mentioned, each surfactant demonstrated highest displacement of oil at a concentration of surfactant in water of 0.002%, with the pluronic surfactant L63 being the most efficient in displacing oil in the Hele-Shaw model (Fig. 9-3).

Oil displacement in the Hele-Shaw model was also evaluated in terms of the spreading coefficient (Becker, 1960). For one liquid to spread on the surface of another immiscible liquid, the work of adhesion (W_a) between two liquids must be greater than the work of cohesion (W_c) of a liquid to itself. The spreading coefficient is the difference between W_a and W_c; a functional equation expressing this relationship is:

$$\gamma_{sp} = W_a - W_c = \gamma_B - (\gamma_A + \gamma_{AB}) \tag{9-2}$$

where γ_{sp} is the spreading coefficient of liquid A spreading on liquid B, γ_B is the surface tension of liquid B, γ_A is the surface tension of liquid A, and γ_{AB} is the interfacial tension of the two liquids.

Spreading coefficients and interfacial tensions for various concentrations of surfactants L63, L61 and F87 were evaluated along with oil displacement ability in the Hele-Shaw model (Fig. 9-5). Low concentrations of surfactant produce high spreading coefficients, indicating that the adhesive properties of the two liquids are dominant. Under these conditions the liquids tend to maximize the interface between them. This correlates with fingering of the aqueous front as it projects into the oil. This maximizes the interfacial area between the two liquids and is unfavorable for total oil displacement.

Conditions which favor greater oil displacement are minimal interfacial area between liquids and a wide lateral spread (which produces wider fingers or a single front). As this occurs, the adhesive properties of the liquids are less predominant with respect to the cohesive properties of the oil phase (the spreading coefficient decreases). A relative increase in cohesive properties of the oil phase discourages penetration, or narrow fingering of the aqueous phase into the oil, and encourages lateral spread. This concept is useful in explaining the observed increases in oil displacement from the Hele-Shaw model at the 0.002% concentration of surfactant. At this concentration the spreading coefficient is approximately 0 ± 1 mN/m (W_a and W_c are approximately equal) and this changes very little with increasing surfactant concentration. It has also been observed that with concentrations of surfactant greater than the optimal, oil displacement decreases. This may be due to a critical interfacial tension for fingering (where microfingers are formed) which cannot be observed well with this model. Interfacial tensions below this critical level favor narrow fingering and thus decrease displacement of oil.

A certain volume of oil spreads on an aqueous medium until an equilibrium area is attained. The greater the thickness of the oil, the smaller the spreading area (Fig. 9-6). Thus at defined volumes of oil and water, thickness measurements can be used as an indication of spreading. Thickness measurements were made at specified concentrations of surfactant L63 corresponding to the greatest thickness (smallest droplet area) at the interface. The least amount of spreading was observed at the optimal concentration of surfactant for oil recovery, i.e., 0.002%. This further supports the concept that at this surfactant concentration, cohesive forces of the oil

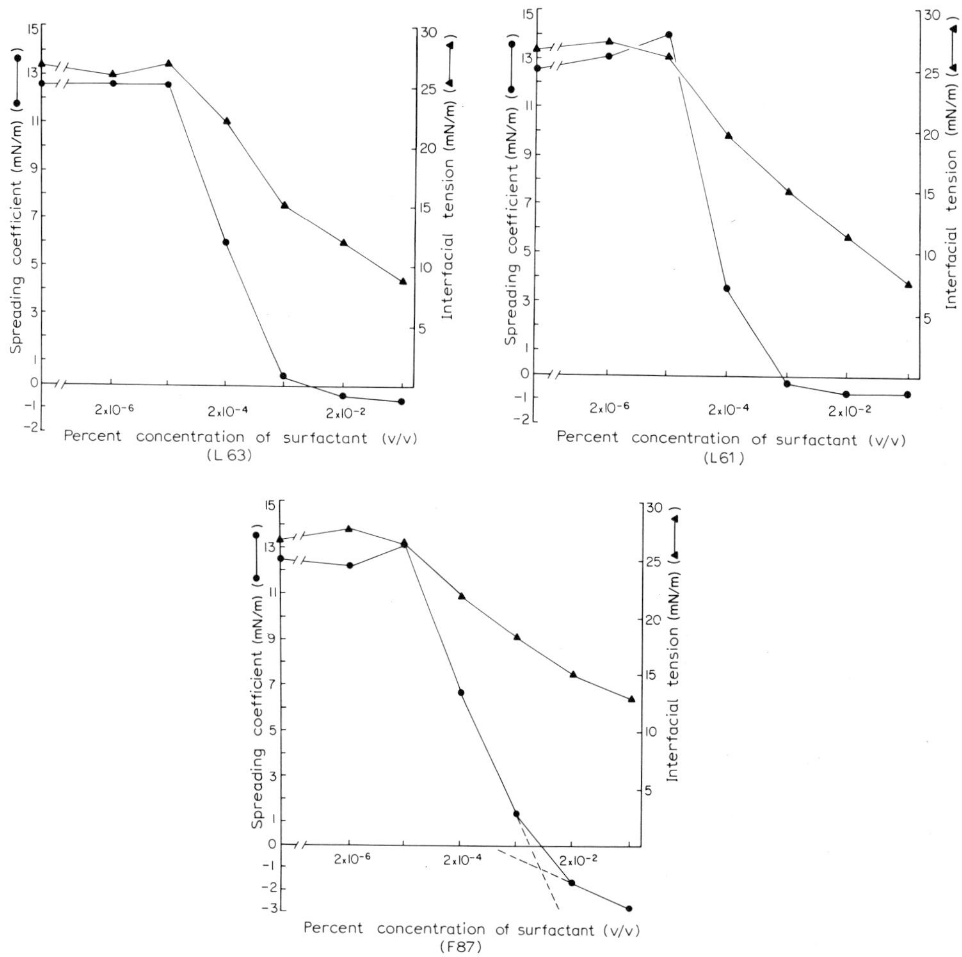

Fig. 9-5. Spreading coefficients and interfacial tensions for pluronic surfactants L63, L61, and F87 at various concentrations.

are predominant as compared to the adhesive properties of the two liquids. Cohesive properties tend to increase the contact angle of the oil–surfactant-water interface, whereas adhesive properties would cause increased spreading at the interface. As fingering effects become minimized and a smooth lateral spreading front along the interface is favored, displacement of oil is increased.

Oil recovery in the Hele-Shaw model by surfactant solutions seems to be dependent on two main factors: spreading coefficient and interfacial tension. A balance between the two seems necessary for optimal displacement of oil. From Fig. 9-5 it appears that a maximum difference between the interfacial tension (γ_i) and the spreading coefficient (γ_{sp}) are desirable. Solutions with low spreading coeffi-

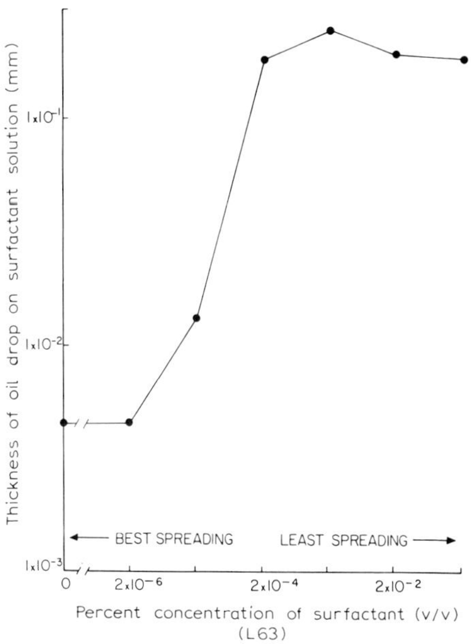

Fig. 9-6. Relationship between the surfactant concentration and height of an oil drop placed on surfactant solutions (a measure of spreading).

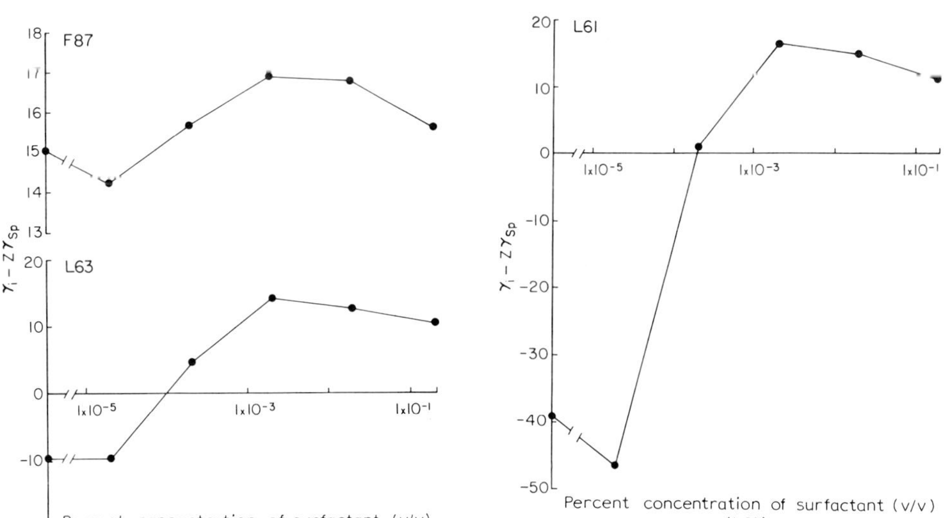

Fig. 9-7. Relationship between surfactant concentration and $(\gamma_i - Z\gamma_{sp})$ for pluronic surfactants F87, L63, and L61.

TABLE 9-I

Z-factor values for three pluronic surfactants used in oil displacement studies from a single pore

Surfactant	HLB [a]	CMC [b]	Z
L61	3	4.5×10^{-4}	5.25 ± 3.95
L63	11	2.9×10^{-4}	$2.9\ \pm1.6$
F87	24	5.1×10^{-4}	0.95 ± 0.05

[a] Hydrophilic–lipophilic balance.

[b] Critical micellar concentrations.

cients (approximately 0) and interfacial tensions of about 15 mN/m appear to be most effective at displacing oil from the Hele-Shaw model.

The difference between interfacial tension and spreading coefficient ($\gamma_i - Z\gamma_{sp}$) may be used to generate curves very similar to those seen for the percent of oil recovered for surfactant solutions (Fig. 9-2). Plots of ($\gamma_i - Z\gamma_{sp}$) vs. surfactant concentration for the surfactants L63, L61 and F87, are shown in Fig. 9-7. A peak is observed at a surfactant concentration of 0.002%, which corresponds to that observed for optimal oil displacement suggesting that this equation may be used to determine optimal γ_i and γ_{sp} values for a surfactant solution. The Z value above is a relative weight value for the importance of γ_{sp} for developing a peak in oil recovery. Table 9-I gives values for Z determined for three surfactants: L61, L63, F87. The values ranged from 0.95 to 5.25. The large values of Z indicate that γ_{sp} contributes more than γ_i for oil displacement from single pores. Table 9-I also gives critical micellar concentrations (CMC) and HLB's for these surfactants (Adamson, 1976). The value of optimal surfactant concentration for oil recovery (0.002%) is greater than the CMC value, indicating that it is a micellar-controlled system.

OIL DISPLACEMENT WITH POLYMER–SURFACTANT SOLUTIONS

Polymer ability to increase the viscosity of solutions is an important factor in oil displacement in porous media (Zajic et al., 1983). A study was undertaken to examine reduction of interfacial tension (using pluronic surfactant) in viscous solutions and its impact on oil displacement from the Hele-Shaw model.

The pluronic surfactant L63 was added in varying concentrations to 0.1% xanflood solutions (viscosity of approximately 44 cP). Results obtained after evaluation for oil displacement ability are presented in Fig. 9-8a. First, it should be explained that polymers increase oil recoveries far greater than water–surfactant solutions in the Hele-Shaw cell. This is attributed to the high viscosity which reduces fingering and gives a broad single front to push oil out of the chamber. Although the oil recoveries are higher, surfactant contribution to oil recovery can be evaluated. The surfactant concentration which allows for highest displacement of oil was 0.006%. The interfacial tension at this concentration (indicated in Fig. 9-8a) was

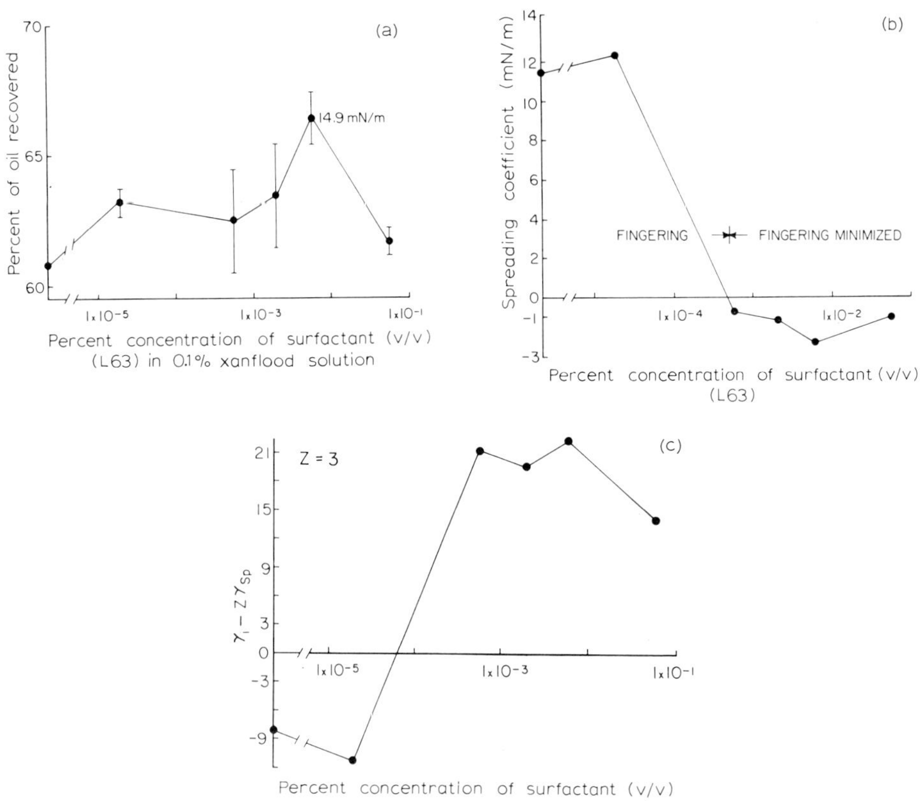

Fig. 9-8. (a) Relationship between the recovery of oil in the Hele-Shaw model using a 0.1% xanflood solution and concentration of pluronic surfactant L63. (b) Relationship between the surfactant concentration and spreading coefficients for the xanflood–surfactant solution. (c) Relationship between the surfactant concentration and $(\gamma_i - Z\gamma_{sp})$ for the xanflood–surfactant system.

TABLE 9-II

Correlation group values of various solutions of the pluronic surfactant L63 tested in the Hele-Shaw model

Concentration of L63	Calculated correlation group (equation (9-1))	Percent oil recovery
0.0	0.36	26.25
0.00012%	0.46	35.75
0.0008%	0.6	38.2
0.002%	0.65	50.8
0.008%	0.80	32.0
0.06%	1.03	32.0

[a] Correlation group value is greater than 1 for concentrations in excess of this value, yet oil recovered remains the same (32%), due to the occurrence of optimum interfacial tension for oil recovery in the Hele-Shaw model.

14.9 mN/m (i.e., close to the same critical interfacial tension reported without a polymer).

Spreading coefficients were evaluated for this system and are presented in Fig. 9-8b. A spreading coefficient of zero was observed at a concentration of 0.005% L63. Oil recovery increased as the spreading coefficient decreased for solutions with interfacial tensions greater than the critical interfacial tension for fingering solutions, whereas interfacial tensions below this decreased oil recovery. Oil recovery was highest with $(\gamma_i - Z\gamma_{sp})$ at 19.5, which occurs when the surfactant concentration reaches 0.004% or greater. The Z value in this case is then greater than or equal to 2.6. Fig. 9-8c shows a $(\gamma_i - Z\gamma_{sp})$ curve obtained using a Z value of 3.

Thus it was observed that the contribution of surfactant solutions may be evaluated in the single pore system from increased oil recovery by polymer solutions. Again, a balance between contributions of interfacial tension and spreading coefficient had to be considered to optimize the displacement of oil (Table 9-II). These results show that manipulation of a displacing fluid both by viscosity control and reduction of interfacial tension is required for high enhanced oil recovery.

EFFECT OF BACTERIAL CULTURES

Undiluted as well as diluted cultures of *Corynebacterium fascians* and *Bacillus subtilis* were examined for microbial enhancement of oil recovery (MEOR) in the Hele-Shaw model. Both of these cultures are known to produce surfactants: the former produces lipid-type compounds and the latter subtilysin. Results obtained using dilutions of a *C. fascians* culture are presented in Fig. 9-9a. Each data point represents a dilution of a *C. fascians* culture, producing different interfacial tensions. Dilutions producing very low interfacial tensions proved to be less effective in displacing oil in the Hele-Shaw model than water (these very low interfacial tensions may be due in part to the hydrocarbon content of the medium).

With interfacial tensions ranging from 4 mN/m to that of water (27 mN/m), all dilutions showed slight improvement over water for oil displacement ability. This indicates that the critical interfacial tension for fingering occurs at approximately 4 mN/m rather than that observed previously at 15 mN/m. The occurrence of a lower critical interfacial tension may be due to a very low spreading coefficient produced for this system as shown in Fig. 9-10a (much lower than those seen for the pluronics). Since the spreading coefficient is very low, one can assume that the contribution of the spreading coefficient to W_c of the oil phase is a dominant factor in this system. Cohesive properties of the oil would discourage narrow fingering into the oil and would encourage lateral spreading, thus increasing displacement of oil.

It appears that as the interfacial tension decreases, oil displacement increases. The line in Fig. 9-9a indicates increasing oil recovery with decreasing γ_i of *C. fascians* broth. The correlation coefficient for this line is 0.544, giving an approximately 90% confidence in linearity. The interfacial tension decreases because of a

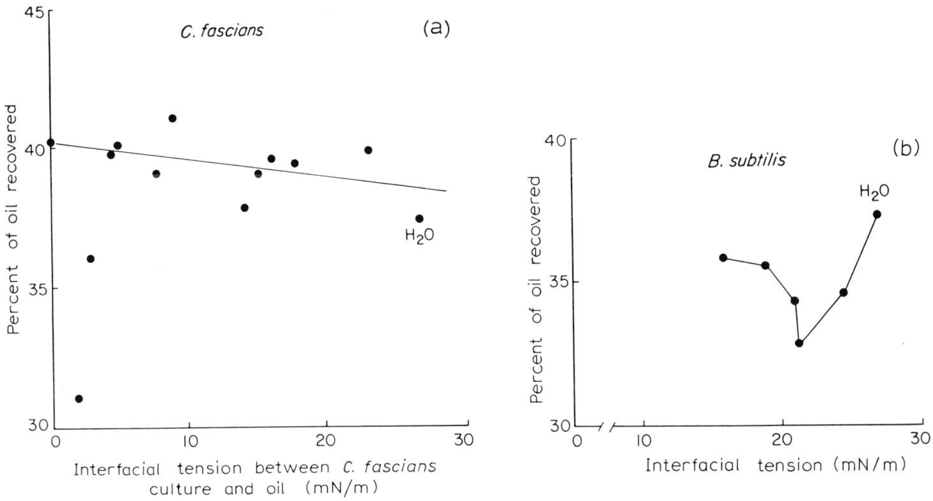

Fig. 9-9. (a) Relationship between the recovery of oil in the Hele-Shaw model and interfacial tension between the *C. fascians* culture and oil. (b) Relationship between the recovery of oil in the Hele-Shaw model and interfacial tension using a *B. subtilis* culture.

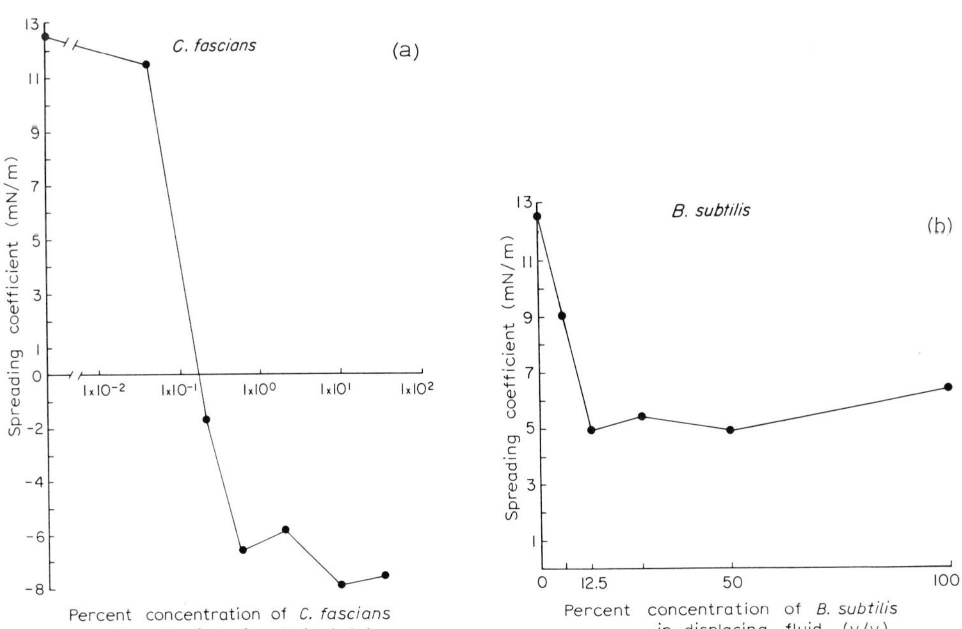

Fig. 9-10. Relationship between spreading coefficient and concentration of the *C. fascians* (a) and *B. subtilis* (b) systems in displacing fluid.

greater concentration of *C. fascians* in the displacing fluid. From Fig. 9-10a, a greater concentration actually shows a lower spreading coefficient. Low spreading coefficients produce conditions which favor increased displacement of oil from the Hele-Shaw model. Since the spreading coefficients for *C. fascians* are so much lower than those seen previously, the interfacial tension at which fingering (a condition which favors decreased displacement) begins is also lowered. The enhanced effect of the spreading coefficient reduces the critical interfacial tension for fingering.

Results obtained using dilutions of a *B. subtilis* culture are presented in Fig. 9-9b. The lowest interfacial tension obtainable with this culture was 16 mN/m. The culture was diluted to produce various interfacial tensions up to that of water. All dilutions of the *B. subtilis* culture failed to perform as well as water in the Hele-Shaw model. The spreading coefficients produced by the various dilutions of *C. fascians* and *B. subtilis* are presented in Fig. 9-10a, b. Low spreading coefficient values observed in other systems were never reached, indicating conditions less favorable for oil displacement in the Hele-Shaw model.

C. fascians appears to have some potential in MEOR techniques from the Hele-Shaw model. The strain of *B. subtilis* used in this test is of limited interest. Other surfactant-producing cultures must be examined.

CORRELATION GROUPS

Equation (9-1) represents the correlation group which can be used for evaluation of oil recovery in the Hele-Shaw model. Mobility control also may be used to evaluate oil recovery, but this parameter is reserved primarily for viscous solutions (Mungan, 1971). The $(\gamma_i - Z\gamma_{sp})$ term, which describes the effects of interfacial tension and spreading coefficient, appears to be an important functional parameter for the evaluation of oil displacement from the Hele-Shaw model by surfactant solutions.

The correlation coefficient (equation (9-1)) describes the shape of the meniscus formed at the interface between water and oil, when water is displacing oil from the Hele-Shaw model (Saffman and Taylor, 1958). The shape of the meniscus influences the amount of oil displaced: the larger the meniscus or the wider the finger, the more oil is displaced. Favorable conditions for oil displacement in the single-pore model are high flow rates, high viscosity, and low interfacial tensions. In large pore systems, ultra-low γ_i values are not required. This indicates that large values for the correlation coefficient are desirable. For purpose of the studies herein, interfacial tension is the predominant factor in the correlation coefficient, followed by the flow rate, particularly when viscosity is held constant for the different solutions tested (i.e., is equal to that of water).

Examining the correlation group values and amount of oil recovered for solutions of different surfactants (Table 9-III), each at a concentration of 0.002%, it was observed that the maximum oil recovery is again obtained with an intermediate

TABLE 9-III
Correlation group values of various surfactant solutions all at a 0.002% concentration (arranged in order of increasing interfacial tension)

Surfactant (0.002% conc.)	Correlation group value (equation (9-1))	Percent oil recovered
P75	0.69	42.5
L62	0.67	43.3
L63	0.65	50.8
L61	0.62	48.8
F87	0.62	44.8
F38	0.52	38.3
H_2O	0.36	37.3
0.19 xanflood w/0.006% L63 [a]	2.33	66.6

[a] Viscosity of this polymer solution becomes a predominant factor; a high correlation group value corresponds to a high percentage of oil recovered.

correlation coefficient value. This is caused by the occurrence of an optimal concentration as well as a surfactant which optimizes conditions for oil displacement in the Hele-Shaw model. In general, all correlation group values are small due to the low viscosity of displacing fluids.

CONCLUSIONS

Oil recovery in a large single pore (typified by the Hele-Shaw model) by surfactant solutions at a constant viscosity seems to be affected predominantly by two main factors: (1) spreading coefficient, and (2) interfacial tension. There appears to be a balance between the two, produced by an optimal concentration of surfactant, which is desirable to enhance oil recovery in the Hele-Shaw model. Conditions which optimize displacement of oil from the Hele-Shaw model are a low spreading coefficient and a critical interfacial tension as it relates to reduction of fingering. For the pluronic surfactant systems evaluated, the critical interfacial tension is equal to approximately 15 mN/m. For the bacterial systems (027090C027R; *C. fascians*) this value was lower: 4 mN/m.

REFERENCES

Adamson, A.W., 1976. *Physical Chemistry of Surfaces*. John Wiley and Sons, New York, N.Y., 476 pp.

Becker, P., 1960. *Spreading, HLB, and Emulsion Stability*. Society of Cosmetic Chemicals, p. 325.

Gogarty, W. and Fosch, W., 1968. Miscible-type waterflooding: oil recovery with micellar solutions. *J. Pet. Technol.*, 20(12): 1407–1414.

Hele-Shaw, H., 1897. *Trans. Inst. Naval Arch.*, 39: 145.

Hele-Shaw, H., 1898. Investigations of nature of surface resistance of water and of stream-line motion under certain experimental conditions. *Trans. Inst. Naval Arch.*, 40: 21.

Mungan, N., 1971. Improved waterflooding through mobility control. *Can. J. Chem. Eng.*, 49(1): 32–37.

Saffman, P.G. and Taylor, G., 1958. Penetration of a fluid into a porous medium or Hele-Shaw cell containing a more viscous liquid. *Proc. R. Soc. London, Ser. A*, 245: 312.

Scheidegger, D., 1960. *Physics of Flow Through Porous Media*. Toronto Press, Toronto, Ont., pp. 73–98.

Taber, J.J., 1969. Dynamic and static forces required to remove a discontinuous oil phase from porous media containing both oil and water. *Soc. Pet. Eng. J.*, 9(1): 3–12.

Taber, J.J., 1981. Research on enhanced oil recovery: past, present and future. In: D. Shah (Editor), *Surface Phenomena in Enhanced Oil Recovery*. Plenum Press, New York, N.Y., pp. 13–52.

Zajic, J.E., Hostak, J. and Seffens, W., 1983. The effect of HLB on the surface activity and bitumen extraction capability of *Corynebacterium fascians*. In: E.C. Donaldson and J.B. Clark (Editors), *Proceedings, 1982 International Conference on Microbial Enhancement of Oil Recovery*. U.S. Department of Energy, NTIS, Springfield, Va., pp. 15–19.

Chapter 10

BIODEGRADATION OF CHEMICALS USED IN ENHANCED OIL RECOVERY

MARY M. GRULA, ROD PENNINGTON and GUY W. SEWELL

INTRODUCTION

Several technologies for enhanced oil recovery are available (Geffen, 1976), including thermal, non-hydrocarbon gas injection, and chemical flooding. The direct use of microbes (MEOR) is undergoing active development. In the case of chemical flooding, the compounds used are subject to varying degrees of biodegradation and biodeterioration, changes that involve the activities of microbes. Thus, this topic is within the scope of the general area of microbial aspects of enhanced oil recovery.

As thermal and non-hydrocarbon gas injection EOR technologies are not significantly involved with microbes, they will not be considered further here. In the chemical flooding category, the important processes are polymer flooding, surfactant flooding, and alkaline flooding. Since only the first two use organic compounds, they are the only ones considered in this chapter. A combination of surfactant and polymer (micellar–polymer) flooding in which surfactant is used first, followed by flooding with polymer, is said to be the most effective chemical flooding technique, but it is also the most costly (Geffen, 1976).

Surfactants used are generally petroleum sulfonates, formed by the sulfonation of crude oil. This results in a very complex mixture of compounds, in which the aromatic and alkyl-aryl components have been sulfonated on the aromatic ring. Introduction of the ionizable, hydrophilic -SO_3H group gives the mixture the property of forming stable emulsions in water.

The polymers most widely used are polyacrylamides (linear copolymers of acrylamide $CH_2=CH\text{–}C\langle{}^{O}_{NH_2}$ and acrylic acid — $CH_2=CH\langle{}^{O}_{OH}$), in which the ratio of carboxyl groups to amide groups in the residues is called the degree of hydrolysis, and biopolymers (biologically synthesized), of which xanthan gum, a heteropolysaccharide produced by *Xanthomonas campestris* NRRL B-1459, is the most important.

These polymers are very water-soluble, forming viscous solutions which can serve as mobility control agents. Other polymers which have been used to some extent as mobility control agents are carboxymethyl cellulose, hydroxyethyl cellulose, and polyethylene glycol monobutyl ether.

Co-surfactants are generally used in flooding operations involving surfactants. The most commonly used chemical cosurfactants are aliphatic alcohols with four to eight carbon atoms.

In addition to the previously mentioned chemicals, various biocides, chelating agents, and oxygen scavengers are commonly used. These include, but are not necessarily limited to, quaternary ammonium salts, 2,4,5-trichlorophenol, pentachlorophenol, phenol, 2,2-dibromo-3-nitrilopropionamide (DBNPA), copper sulfate, glutaraldehyde, formaldehyde, sodium hypochlorite, acrolein, ethylene diamine tetraacetic acid (EDTA), and 1,6-hexane diamine (Silvestro and Desmarais, 1980).

Because of the potentially very broad scope of this chapter, an attempt was made to maintain a reasonable length by limiting the compounds considered to (1) those for which biodegradability has been a subject of controversy (linear alkyl benzene sulfonates, or LAS); (2) those about which little is known (non-linear alkyl benzene sulfonates, or ABS); and (3) the most widely used types of polymers, viz., polyacrylamides and xanthan gum.

The alcohols used as co-surfactants are for the most part biodegradable, although they vary in degree of biodegradability. Those with branched carbon chains, such as tertiary butyl alcohol, are generally less degradable than those with straight chains.

The question of the biodegradability of each of the very large number of hydrocarbon components of petroleum sulfonates, of co-surfactants, and of types of polymers other than polyacrylamides or xanthum gum are discussed briefly, because an attempt at exhaustive coverage of these materials would require space equivalent to the length of this entire book.

FUNDAMENTALS OF BIODEGRADATION AND BIODEGRADABILITY

Biodegradation is generally recognized as a series of natural processes resulting from the activities of microorganisms, by which organic materials are converted to simpler compounds and finally to inorganic substances (CO_2, CH_4, H_2O, NO_3^-, SO_4^{2-}, PO_4^{3-}, etc.), most of which are utilizable as nutrients by green plants. These processes, in totality, constitute the cycling of the elements in the biosphere, upon which the continuation of life on earth depends. Biodegradability is the state or capability of being degraded by biological processes.

Biodegradation is largely carried out in nature by two groups of organisms, viz., the heterotrophic bacteria and the fungi. Bacteria are prokaryotic organisms, i.e., they do not possess a true nucleus, but rather a "nuclear body" or "nucleoid", a structure unenclosed by a nuclear membrane which consists of a closed circle of double-stranded DNA. This body is also often referred to as the bacterial chromosome. It contains genes with codes for all or most of the traits essential for the continued existence of the bacterial species. In addition to the chromosome, many bacteria (especially when freshly isolated) contain small (about 1–3% the size of the chromosome) circles of DNA called plasmids. These structures contain genes governing traits which are generally not essential for continued life of the cell. It is of interest that the ability to attack and utilize a given organic compound is often a plasmid-determined trait.

In addition to the eubacteria ("true" bacteria), the prokaryotes include the actinomycetes, slow-growing organisms of roughly bacterial dimensions, many of which have a more complex morphology (such as a mycelial structure) than eubacteria. The other group of organisms actively involved in biodegradation, the fungi, are eukaryotic, i.e., they possess a true nucleus bounded by a nuclear membrane. Replication of nuclear DNA and cell division in eukaryotic cells takes place by a process involving chromosome pairing and alignment called mitosis. In general, prokaryotic cells contain only one copy of each gene (haploid state), whereas eukaryotic somatic cells contain two copies (diploid state), or more in some cases.

Biodegradation often (but not always) involves utilization of the substance being broken down as a source of carbon-energy by the degrading organism. Under aerobic conditions, given complete utilization, a portion of the degraded substance will be converted to oxidized products (CO_2, H_2O, oxidized inorganic molecules), and a portion will be converted to cell material.

Many chemicals are broken down by biological process in which the compound does not serve as a source of carbon-energy for the microbes carrying out the degradation. This is especially true of xenobiotic compounds, i.e., compounds which have been synthesized in the laboratory, but not by a natural process. Decomposition proceeds in a slow and linear fashion, and is not accompanied by any increase in population or biomass. Often the compound is only partially decomposed, rather than being converted to inorganic materials and biomass. The first clue that this type of breakdown is a biological process was the fact that it occurred in non-sterile soil, but not in sterile soil (Alexander, 1981). Reactions characterized in these ways are termed co-metabolism or co-oxidation (Alexander, 1981). Some compounds of low biodegradability may be broken down only by co-metabolic reactions; in these cases, no organism able to use the compound as a sole source of carbon has ever been isolated.

Many examples of degradation via co-metabolic reactions are known. In general, chlorinated organic molecules, including DDT, 2,4,5-T, aldrin, and heptachlor, are transformed via co-metabolic reaction (Leisinger et al., 1981). It is possible that compounds once thought to be transformed biologically only by co-metabolic reactions will someday be shown to be assimilable by a microorganism, most probably as a carbon-energy source. For example, Edgehill and Finn (1983) reported isolation of a strain of *Arthrobacter* (ATCC 33790) that is capable of utilizing pentachlorophenol (PCP), a compound that a priori would appear to be quite recalcitrant.

Other workers (Stanlake and Finn, 1982; Watanabe, 1973; Chu and Kirsch, 1972) have also isolated bacteria capable of degrading pentachlorophenol. Stanlake and Finn (1982), who isolated several *Arthrobacter* strains capable of utilizing PCP as a sole carbon source, state that PCP is a recalcitrant molecule capable of being degraded by only a limited number of bacteria. In many environments PCP would probably prove to be quite persistent.

Degradation of highly chlorinated biphenyls by bacteria can be considered to be a cometabolic reaction (Leisinger et al., 1981). However, a biphenyl with only one chlorine substituent can serve as a sole source of carbon in a mineral medium for an *Acinetobacter* species. It is possible that many compounds thought to be transformed only by co-metabolic reactions could be shown to be metabolized if adequate methods are used to find the right organism.

It would be well to distinguish at this point between biodegradation and biodeterioration. Rose (1981) defined biodeterioration as a change in the quality or value of a material by the activities of microorganisms, which makes the material less useful in aesthetic or utilitarian terms. Thus, the anaerobic corrosion of metal pipes by the action of sulfate-reducing bacteria should be viewed as a form of biodeterioration rather than biodegradation. It should be apparent that a sharp distinction between these two processes is not always possible. The effects of microbial activities on polyacrylamides as observed by oilfield workers might more properly be considered biodeterioration rather than biodegradation.

FACTORS AFFECTING BIODEGRADATION AND RATES OF DEGRADATION

The rate at which a given compound will degrade when placed in a certain natural environment is determined by many factors which interact in a complex manner. Some of these reactions will be non-biological in nature. Only few chemicals are completely non-reactive in the environment, i.e., their concentrations change only as a result of the physical processes of dilution and dispersion (Maki et al., 1980). Most chemicals undergo a variety of transformations (particularly in an aqueous environment), such as acid–base, complexation, oxidation–reduction, sorption (including biological uptake), hydrolysis, photolysis, and biotransformation (Maki et al., 1980).

With one group of chemicals used in enhanced oil recovery, viz., the alkyl benzene sulfonate surfactants, the paramount importance of biodegradation as opposed to purely chemical degradation is indicated by the fact that non-linear alkyl benzene sulfonates (ABS) degrade very slowly in aquatic environments. This resulted in serious problems of foaming in sewage treatment plants and even in water supplies when ABS came into widespread use as a component of laundry detergents, in place of soap, roughly in the time period 1950–1966 (Nielsen and Huddleston, 1981). With the introduction of the biodegradable LAS, foaming in water supplies has largely disappeared. This experience may be regarded as a large-scale field test illustrating differences in degree of biodegradability. It seems reasonable that LAS and ABS would undergo similar types of *chemical* (abiotic) reactions at approximately the same rate, yet their rates of biodegradation are very different.

Apart from the inherent biodegradability of a compound, many factors will affect the rate at which biodegradation of a given compound takes place in a certain environment. The first essential factor is the presence in the environment of a population of microorganisms with the appropriate catabolic mechanisms for degrading the compound in question (Slater and Somerville, 1979). The rate of biodegradation depends on the initial interaction between the compound and the organism, the concentration of the degrading population and the compound concentration, and the physicochemical parameters of the environment (temperature, pH, oxidation potential, salinity, nutrient content, presence of toxic materials, etc.) (Slater and Somerville, 1979). The concentration of the chemical being degraded is very important. Many compounds which are inherently biodegradable become toxic for most bacteria if the concentration becomes too high (phenol is an example).

In an actual environment, concentration of a chemical at any given point is a function of many interacting variables, and prediction of the rate of transformation of a chemical in that environment is a very complex matter. Many ways have been devised for investigating microbial degradation in soil and in water (Howard et al., 1975; Bourquin and Pritchard, 1979). The most frequently used tests involve river die-away, biochemical oxygen demand, model ecosystems, microcosms, soil enrichment cultures, and pure cultures. A useful evaluation of methods for determining biodegradability and rates of degradation under various conditions, use of modeling for predictive purposes, and other aspects of environmental fate of polluting chemicals is given in the published proceedings of a workshop on this subject (Bourquin and Pritchard, 1979).

BIODEGRADABILITY OF CHEMICALS USED IN ENHANCED OIL RECOVERY

Water soluble polymers

As stated previously, the polymers most widely used in enhanced oil recovery are polyacrylamides produced by a chemical (non-biological) polymerization, and the biopolymer xanthan gum, produced by the metabolic reactions of a microorganism, *Xanthomonas campestris*. Each type of polymer has certain advantages and disadvantages for use in enhanced oil recovery. Xanthan gum produces a high viscosity solution even at low gum concentrations. In the presence of 0.1% sodium chloride, the viscosity is virtually unaffected by temperatures ranging from 25° to 200°F (−4 to 150°C). The viscosity is also unaffected by pH; at the same salt concentration, it is independent of pH in the range 1.5–13 (Kang and Cottrell, 1979). It is widely perceived as being readily biodegraded, however, under both aerobic and anaerobic conditions (A. Laskin, personal communication, 1984). Polyacrylamides, on the other hand, are regarded as more resistant to biodegradation, but they are very sensitive to shear degradation (Maerker, 1975; Jennings et al., 1976).

Polyacrylamides are produced by a strictly chemical polymerization, and are not known to be synthesized by any organism. As such, they are classified as “xenobiotic”, i.e., strangers to the biosphere. An excellent monograph on the general principles of microbial degradation of xenobiotics and recalcitrant compounds is available (Leisinger et al., 1981). Biodeterioration of plastics (polymers, not highly elastic or crystalline, formulated with a variety of additives to enable fabrication into useful products) is covered in a review by Klausmeier and Andrews (1981).

The backbone of polyacrylamide polymers consists of –CH– groups linked by covalent bonds; thus it is not subject to hydrolytic cleavage as are many biopolymers (starch, cellulose, proteins, etc.). The size of the molecule (molecular weights in the range 10^6–10^7 daltons) should render biodegradability of polyacrylamides unlikely for two reasons: (1) the longest $(-CH_2-)_n$ chain subject to attack by hydrocarbon-degrading microorganisms is approximately only a little over 40 carbons in length; and (2) the molecular size precludes passage of the molecule through the cell membrane—no transport systems or extracellular enzymes attacking polyacrylamides are known to exist.

Yet, the belief appears to persist that somehow polyacrylamides are subject to some kind of microbial attack *. Enrichment cultures of soil microorganisms with polyacrylamide added as a sole carbon source provide evidence that the presence of polyacrylamide in a soil environment stimulates the growth of microorganisms (Grula and Grula, 1983). The mechanism of this stimulation has not been fully elucidated, but one possible explanation is that the release of NH_4^+ ion through hydrolysis of the amide groups of the polymer provides the bacterial population with another source of nitrogen (Grula and Huang, 1981). Most common soil bacteria utilize ammonia, or ammonium salts, as their preferred nitrogen source. Grula and Huang (1981) showed that certain polyacrylamides, of varying molecular weight and degree of hydrolysis, could serve as a sole nitrogen source, in a chemically defined medium, for several soil pseudomonads. All of these strains resembled *Pseudomonas aeruginosa*, and had been isolated from soil enrichment cultures containing polyacrylamide. Growth of one strain, in a defined medium, as a function of polyacrylamide concentration with three different carbon sources, is shown in Fig. 10-1.

It was further observed by these authors that: (1) degree of hydrolysis was not a factor in determining extent or growth; (2) different strains responded to polyacrylamides differently—some isolates, although they could utilize ammonium chloride, did not grow with polyacrylamide as a nitrogen source; and (3) ability of a sugar to allow growth varied with the type of polyacrylamide used as a nitrogen source. These observations indicated that the polyacrylamides used supported growth, not simply because of their residual ammonium ion content, but also (at least in part) because the growing bacteria possessed an enzyme(s) (amidase?) that

* That this is the case is indicated by conversations with many people associated with the oil industry. This is true even though they are seen as more resistant to biodegradation than biopolymers.

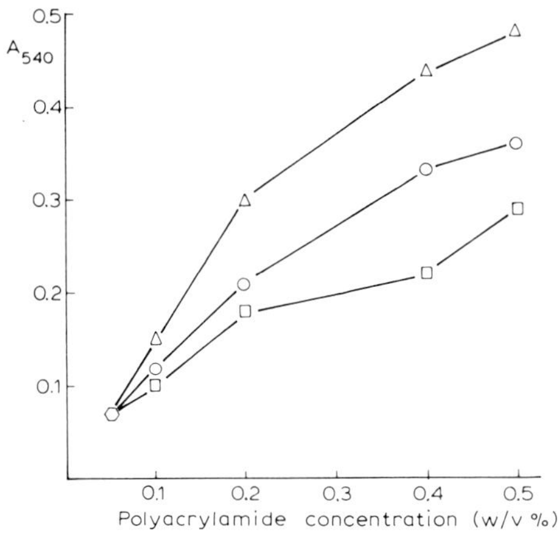

Fig. 10-1. Growth of pseudomonad strain AA as a function of polyacrylamide J279 (1–4% hydrolyzed) concentration. Triangles: glucose; circles: acetate; squares: xylose. Carbon source concentration, 0.1%. A_{540} is absorbance at 540 nm.

enabled them to attack the polymer. An enzyme capable of releasing ammonia from butyramide was demonstrated in one strain of *Pseudomonas* that grew well on polyacrylamide. However, no direct evidence for the existence of amidases capable of releasing ammonia from polyacrylamides was ever obtained.

It was also shown, using a sample of ^{14}C-labelled polyacrylamide J334 (molecular weight = $3–4 \times 10^6$ daltons; 25–35% hydrolyzed) as a sole nitrogen source in a chemically defined medium, that only 0.18% of the total amount of radioactive carbon in the medium was incorporated into the membrane fraction of the cells.

Finally, Grula and Huang (1981) were never able to show any loss of screen factor *, or viscosity of polyacrylamide solutions when the aerobic pseudomonads grew in their presence.

Grula and Sewell (1983) observed that a strain of *Pseudomonas* isolated from a mixture of polyacrylamide and sand (mixture was a sample obtained from an oilfield in Illinois) was markedly stimulated in growth in a chemically defined medium during the early phases of the culture cycle by the presence of polyacrylamide (0.05–0.5%). Ultimate growth attained was not increased by the presence of the polymer. This effect could be partially duplicated by the monomer acrylamide (Table 10-I). After several transfers, and storage, on an artificial medium (trypticase–soy agar), however, this strain lost the property of responding to polyacrylamide. More studies, particularly with freshly isolated strains, must be done before this type of phenomenon can be understood.

* Screen factor of a polyacrylamide solution is a measure of its viscosity and flow properties. It is related to, but not identical with, absolute viscosity. For a discussion of screen factor and its measurement see Jennings et al. (1976).

TABLE 10-I

Stimulation of early growth by polyacrylamide of a pseudomonad isolated from a polymer–sand mixture from an oilfield in Illinois

Medium	Absorbance (O.D.) 540 nm		
	6 hr [a]	7.5 hr	9 hr
Control (no polymer)	0.21	0.48	0.82
0.1% J279 [b]	0.37	0.83	1.20
0.4% J279	0.48	0.96	1.15
1.0 mg/liter acrylamide	0.29	0.73	1.08

[a] Times after inoculation.

[b] M.W. = 3×10^6, 1–4% hydrolyzed. Polymer solution was dialyzed to remove ammonia.

Sample from which this strain was isolated was obtained from an oil reservoir where a polymer flood was in progress.

The interactions of polyacrylamides with anaerobic microorganisms are possibly of greater significance in the recovery of oil using polyacrylamides than the aerobic interaction. Grula and Sewell (1983) observed that crude oil and water samples taken from wells in a portion of a reservoir that had undergone a polyacrylamide flood contained sulfate-reducing bacteria, whereas crude oil and water samples from wells in different regions of the field where there had not been a polyacrylamide flood, had few or no sulfate-reducing bacteria. Using Postgate's Medium E (Postgate, 1979), the sulfate-reducing bacteria were enumerated in the water samples by a 5-tube Most Probable Number (MPN) test. Results are given in Table 10-II.

Informal reports from workers in the oil industry appear to indicate that this phenomenon (increase in numbers of sulfate-reducing bacteria with injection of

TABLE 10-II

Total Most Probable Number (MPN) counts of sulfate-reducing bacteria in water samples from oil wells in the Wilmington Field [a]

Well [b]	MPN/ml produced water
T-100	4.9×10^2
T-101	2.2×10^6
T-102	7.9×10^6
T-103	4.9×10^5
T-104	3.5×10^4
T-105	4.9×10^3
M 486	0
M 526	0
M 711E	24

[a] Located near Long Beach, California. Before the polymer flood, this field had been flooded with sea water.

[b] The T-series of wells had undergone a polymer flood; the M-series had not.

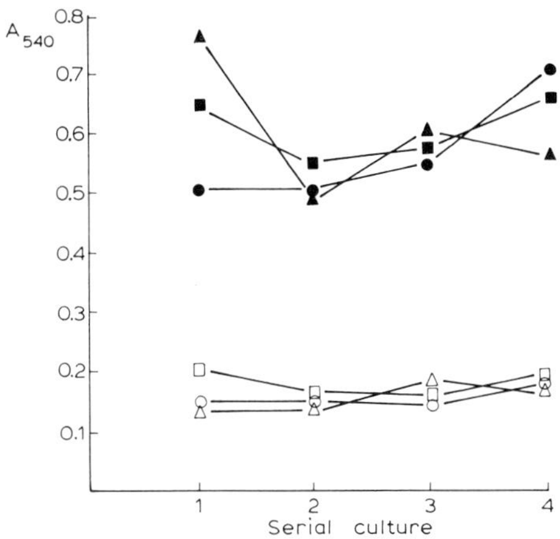

Fig. 10-2. Stimulation of growth of mixed cultures of *Desulfovibrio* by polyacrylamide in the Allred (1954) medium. Circles: culture 2; squares: culture 4; triangles: culture 2+4. Open symbols, without polymer; closed symbols, with polymer. A_{540} is absorbance at 540 nm. These cultures were serially transferred every 48 hours, and were incubated in an anaerobic glove box at 37°C.

polyacrylamide into a reservoir) is fairly widespread. Probably much of the information on which this observation is based is unpublished. More research is needed.

Crude oil samples from the Wilmingtom field in California yielded similar results. Two samples from the T-series of wells (polymer-flooded), when cultured in the Allred medium (Allred et al., 1954) showed sulfate-reducing bacteria, whereas two other samples, from wells in parts of the field that had not been polymer-flooded, did not contain sulfate reducers. An observation on only one field is inadequate to demonstrate a causal relationship between the presence of polyacrylamide solution in an oil reservoir and the occurrence of large numbers of sulfate-reducing bacteria. It would be necessary to have data from many fields, in regions widely separated geographically, in order to rigorously document such a relationship through an empirical approach.

There is no a priori reason why high counts of sulfate-reducing bacteria should be associated with the presence of polyacrylamides. It is extremely unlikely that these polymers could be utilized directly by sulfate-reducing bacteria, either as a source of carbon-energy, or as a source of nitrogen.

Grula and Sewell (1983) noted that certain polyacrylamides stimulated the growth of freshly isolated * sulfate-reducing bacteria under certain conditions (Fig. 10-2).

Some polymers were more effective than others (Fig. 10-3). It is seen that sheared polymer (solution mechanically agitated to break large molecules into smaller units)

* From Wilmington Field water.

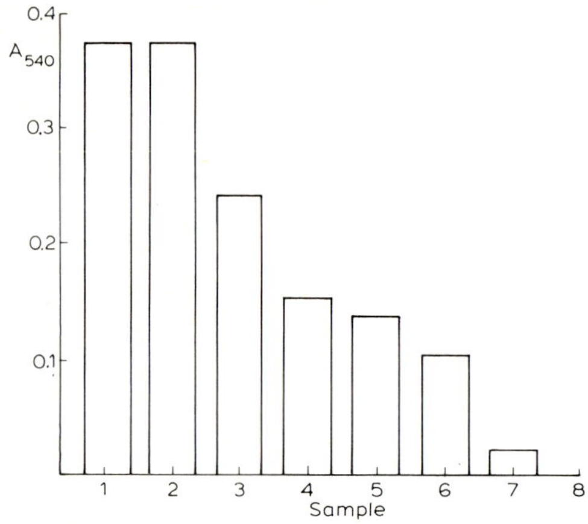

Fig. 10-3. Growth of *Desulfovibrio desulfuricans* strain SB-2-A in Medium C (Postgate, 1979) in the presence of various polymers. Bars: *1* = polyacrylamide J332 (25–35% hydrolyzed, M.W. $9–10 \times 10^6$); *2* = polyacrylamide J332 (sheared); *3* = xanthan gum; *4* = polyacrylamide J279; *5* = control (no polymer); *6* = polyacrylamide (no hydrolysis); *7* = polyacrylic acid; *8* = cationic polyacrylamide. A_{540} is absorbance at 540 nm.

was just as effective as unsheared. Degree of hydrolysis may have been a factor in the effectiveness of growth stimulation; J279 (1–4% hydrolyzed) stimulated much less than J332 (25–30% hydrolyzed).

To a certain extent, the stimulatory effect appeared to be related to the existence of a sulfate deficiency in the medium. The Allred medium, in which the stimulation was first observed (Fig. 10-2), was found to be very deficient in sulfate; it contained

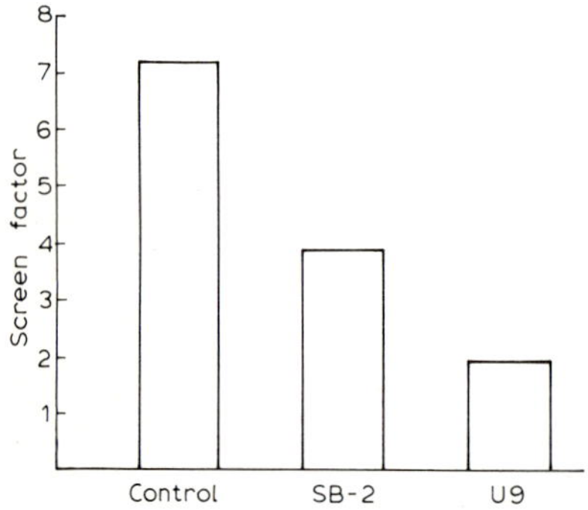

Fig. 10-4. Reduction in screen factor (Jennings et al., 1976) in Postgate's (1979) Medium C by two strains of *Desulfovibrio* isolated from oil reservoir involved in polyacrylamide floods. Polymer used was J332 at 0.1%. Control had no polymer added. Strain SB-2 was isolated from a field near Long Beach, California, and strain U9 from a field near Robinson, Illinois. Incubation was in an anaerobic glove box for 5 days at 37 °C.

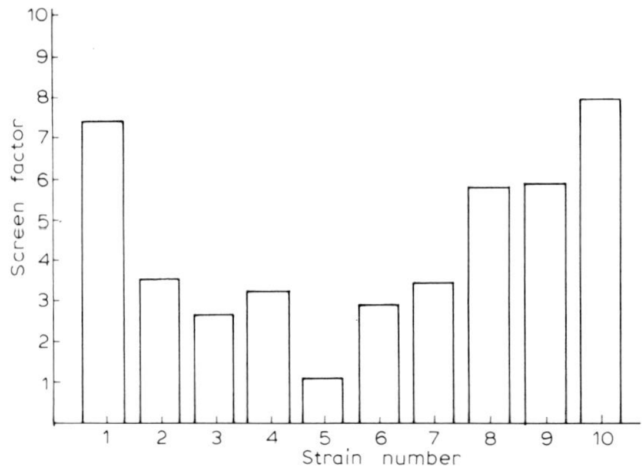

Fig. 10-5. Reduction in screen factor (Jennings et al., 1976) in Postgate's (1979) Medium C by several strains of *Desulfovibrio*. Bars: *1* = control (uninoculated); *2* = SB-2-A; *3* = American Type Culture Collection 7755; *4* = U9; *5* = Medium C with no polymer; *6* = SB 2; *7* = strain B; *8* = strain Am; *9* = strain C; *10* = control (uninoculated). Controls were incubated and tested under the same conditions as the cultures. Incubation was in an anaerobic glove box for 7 days at 37°C. Strains B, Am, and C were from an oilfield in Texas, U.S.A.

only 0.2 g Na_2SO_4/liter, whereas the optimum concentration in Postgate's Medium C was found to be approximately 4 g/liter.

Growth of sulfate-reducing bacteria in the presence of polyacrylamide polymers was found to reduce the screen factor and absolute viscosity (centipoises) of the medium (Figs. 10-4 and 10-5). The mechanism of this reduction is unknown. If it should occur under the conditions of an oil reservoir, the potential damage caused by any extensive proliferation of these organisms during a polymer flood could be very great.

Biodegradation of xanthan gum

Xanthan gum is undoubtedly the most widely used biopolymer in the oil industry. It is used not only as a mobility control agent in enhanced oil recovery, but also to viscosify and control the rheological properties of oilwell drilling muds and to impart high viscosity at low concentrations to fracturing fluids. Its pseudoplasticity (reversible shear-thinning property), excellent compatibility with salts, and resistance to thermal degradation and changes in pH make xanthan gum ideal for these uses by the oil industry (Kang and Cottrell, 1979). In addition, because of its unique rheological properties, xanthan gum finds many uses in other industries, particularly the food industry (Kang and Cottrell, 1979). One example of its use in foods is in diet salad dressings, in which it is the non-caloric thickener and emulsifying agent.

Kang and Cottrell (1979) summarized well the history, safety and regulatory status of xanthan gum, its production by fermentation, and its chemical composition, structure, and conformation. Xanthan gum is a heteropolysaccharide synthesized by *Xanthomonas campestris* NRRL B1459. It has been produced on a commercial basis since 1964.

Xanthan gum is a high-molecular-weight polysaccharide containing D-glucose, D-mannose, and D-glucuronic acid in the molar ratios 2.8 : 3.0 : 2.0, respectively. Kang and Cottrell (1979) summarized information on its primary structure based on the work of Jansson et al. (1975). Xanthan gum is composed of pentasaccharide repeating units containing two glucose residues, two mannose residues, and one glucuronic acid residue. The main chain contains β-D-glucose units linked through 1- and 4-positions, i.e., the chemical structure of the xanthan gum backbone is identical with that of cellulose. The two mannose units and the glucuronic acid unit make up the side chain, which is linked to the 3-position of every other glucose unit, on an average, in the main chain.

The terminal β-D-mannose in the side chain is linked glycosidically to the 4-position of the β-D-glucuronic acid, which in turn is glycosidically linked to the 2-position of the proximal α-D-mannose. The α-D-mannose adjacent to the main chain contains an acetyl group at position 6, and about half the terminal β-D-mannose units are ketalically linked through the 4- and 6-positions to pyruvic acid. The distribution of pyruvate groups has not been fully elucidated. According to Slodki and Cadmus (1983), on an average every other mannose end group has a pyruvic acetal substituent. Every mannosyl residue adjacent to the main chain bears an acetyl group, whereas, in contrast, the number of pyruvic acetal groups on the terminal β-D-mannosyl residue can vary, depending on the strain and on conditions of fermentation. Generally, a lowered pyruvate content is associated with a gum of inferior quality. Sandford et al. (1977) demonstrated the relationship between lowered viscosities of dilute xanthan gum solutions and decreased pyruvic acid content (2.5–3.0%, as compared to 4.0–4.8% in gum of high quality). Dispersions (1.0%) of atypical xanthan gum also failed to give the large increase in viscosity usually observed between 50° and 70°C in the absence of salt (Kang and Cottrell, 1979).

Properties of xanthan gum indicate that it has a specific conformation in solution. Much research has been done recently to elucidate the secondary and tertiary structures of xanthan gum (Cadmus and Slodki, 1984). The molecule appears to be a single-stranded or multi-stranded helix. The nature of the association between chains and the role of the side chain are very important in determining the rheological properties of solutions of the gum. Cadmus and Slodki (1984) stated that the reversible shear thinning occurs because the association between chains is easily broken by shear forces, and easily restored when the shear forces stop.

Xanthan gum is biodegradable aerobically, although, because of the stability of unsterile solutions of the polymer, Cadmus et al. (1982) concluded that xanthan-degrading microorganisms are not abundant in natural environments. By the use of enrichment cultures (xanthan gum as a sole carbon source), these authors isolated a pure culture of a salt-tolerant (4% NaCl) *Bacillus* sp. that produced a salt-tolerant degradative enzyme attacking xanthan gum. The source material was soil collected from the inside of a decaying tree trunk. Another xanthanase-producing *Bacillus* (not salt-tolerant) was isolated from sewage sludge. Enzyme preparations from these organisms bring about a dramatic loss in viscosity of xanthan gum dispersions (e.g., a 1% dispersion is reduced in viscosity from 1750 mPa s to 10 mPa s in 18 hours at 37°C). The enzyme preparations (it is believed that the "xanthanase" preparations are actually mixtures of enzymes) remove the side chains, and produce reducing sugar. The yield of reducing sugar indicates a 38% conversion. The backbone is evidently not attacked, because a residual high-molecular-weight polysaccharide precititable by ethanol but resistant to further attack by the enzyme preparation, remains. It is a D-glucan, and was shown by methylation–fragmentation studies to be the β-1,4-linked cellulose-like backbone.

Cadmus et al. (1982) described in some detail their methods of isolating the xanthanase-producing strains, their properties, methods of culturing for enzymes, enzyme recovery purification, and assay, and analytical procedures.

It is interesting that the salt-tolerant strain (designated K-11) produced much more enzyme when grown in mixed culture with various other bacteria isolated at the same time from the same source (a decaying tree trunk). Even some previously isolated and identified microorganisms, belonging to various genera commonly found in soil, elicited an enhancement of xanthanase production by strain K-11.

"The presence of 2,3,5-tri-O-methyl-D-glucose in the high-molecular-weight fraction, the lack of glucose in the low-molecular-weight fraction, the absence of 2,6-di-O-methyl-D-glucose in the permethylated high-molecular-weight fraction, and the composition of the low-molecular-weight fraction indicate considerable cleavage of side chains from the cellulosic backbone, and their fragmentation. From these results, it appears that the xanthanase was a mixture of enzymes that attack all of the side chain linkages in xanthan gum, including the one involving (1 → 3)-linkage of acetylated mannose to the glucosidic backbone. Because the β-(1 → 4)-linked backbone remained intact, an endocellulose type activity is probably absent." (Cadmus et al., 1982, p. 11.)

It should perhaps be restated at this point that xanthan gum, while it is biodegradable, is relatively resistant to biodegradation. This opinion has been expressed not only by Cadmus and his coworkers, but also by Holzworth (1984). Cadmus et al. (1982), in addition to strain K-11, isolated 21 other strains from the mixed culture from the same source. None of these exhibited xanthanase activity in pure culture.

Xanthan gum is also said to be biodegraded under anaerobic conditions (A. Laskin, personal communication, 1984). Little is known about (1) the nature of the microorganisms bringing about anaerobic degradation of xanthan gum, (2) abundance of these microorganisms in nature, (3) properties of the microorganisms, or (4) the type of reaction on xanthan gum these microorganisms bring about. It is assumed that the degradation is accompanied by a loss in viscosity, because a viscosity loss is the method most commonly used to detect a breakdown of the intact xanthan gum molecule.

Petroleum sulfonates and alkyl–aryl sulfonates

Petroleum sulfonates, produced by the sulfonation of crude oil, are used by the oil industry in chemical flooding enhanced oil recovery processes. These materials are surface-active agents or surfactants, because they possess both a hydrophilic group (sulfonic acid) and a hydrophobic group (the alkyl or aryl hydrocarbon moiety) on the same molecule. They lower the surface tension of aqueous solutions, and, when used in a chemical flooding process in enhanced oil recovery, help to remove the oil from the rock pores and move it toward the wellbore. Petroleum sulfonates are produced by the sulfonation of crude oil, yielding a material that somewhat resembles oil in appearance, but is water-soluble, or at least capable of forming very stable emulsions in water. Petroleum sulfonates are complex mixtures of many chemical compounds.

Alkyl aryl sulfonates, on the other hand, are specific compounds in which an alkyl group and a sulfonic acid group are linked to an aromatic ring (usually but not always benzene). Alkyl benzene sulfonates, in particular, have found very extensive use as laundry detergents since World War II. The alkyl benzene sulfonates belong to the anionic class of surface-active agents, because the organic ion carries a negative charge.

Since modern formulations of surfactants rarely resist biodegradation, Cain (1981) raised the question of how competent microflora of soils and material waters have acquired the ability to break down these xenobiotic materials. This author pointed out that in the case of alkyl sulfates and linear alkyl benzene sulfonates (alkyl group unbranched), the available evidence suggests that the organisms use their enzymes for lipid catabolism to effect ω- and β-oxidation (in some rare cases, α-oxidation) of the alkyl chains. There is also recent evidence to support the view that desulfonation of the aromatic ring in aryl sulfonates may be achieved by modified aryl dioxygenases which normally attack an unsubstituted aromatic nucleus. Cain further stated that such an attack by aryl dioxygenases probably occurs with linear alkyl benzene sulfonates (LAS) also.

The alkyl benzene sulfonates (both linear and non-linear) are undoubtedly components of petroleum sulfonates.

Past production data indicate that, in EOR technologies, far more petroleum sulfonates are used than the relatively pure alkyl aryl sulfonates. For example, Silvestro and Desmarais (1980) gave production figures for 1966 as 596,416,000 lb for petroleum sulfonates, and 8,277,000 lb for alkyl–aryl sulfonates.

Very little information is available on the biodegradation of petroleum sulfonates. On the basis of work with other materials, it would be expected that most of the hydrocarbon components of petroleum sulfonates are biodegradable, some more slowly than others. It is known that aliphatic hydrocarbons are assimilated by a wide variety of microorganisms (Ratledge, 1978). Aromatic hydrocarbons may be oxidized by many microorganisms, but are assimilated by only a few bacteria. This difference could be critical in determining the rate of breakdown in an environmental context, because those substances which are actually assimilated by microorganisms contribute to their nutrition, and generally help to increase the population, whereas those compounds that are oxidized but not utilized (co-metabolism) do not encourage growth or a population increase. Increased numbers of the degrading population will result in an increased rate of degradation, provided no other factors are limiting the degradation process.

Some data are available on the rate of degradation of the various broad classes of components of refinery wastes from the landfarm experiments of Conoco, Inc. Twenty-five months after applying refinery waste (low load) to a plot of ground, the average percent degradation of the paraffins was 71, of the aromatics, 58, and of the resins and asphaltenes, 37. The latter fraction did not show any degradation at 12 months, but did show 35% degradation at 18 months; thus, it has a lower degree of biodegradability than either the paraffins or the aromatics (Meyers and Huddleston, 1979). These findings are in accord with the experience of many other workers in this area. While petroleum sulfonates are certainly not the same as refinery wastes, it would be expected that the hydrocarbon components of petroleum sulfonates are degradable, and that the degree of biodegradability of various types would be of the same order as in refinery wastes.

In considering the biodegradation of aliphatic hydrocarbons, it is useful to keep the following generalizations in mind (Ratledge, 1978, p. 3): "(1) *n*-alkanes of chain length shorter than *n*-nonane are not usually assimilated but may be oxidized. Only some bacteria have the ability to grow on alkanes shorter than *n*-octane. As the chain length of the alkane increases beyond C_9, the yield factor increases but the rate of oxidation decreases; (2) saturated compounds are degraded more readily than unsaturated ones; (3) branched chain compounds are degraded less readily than straight-chain ones." In addition, alkanes of high enough molecular weight to be solids at ordinary temperatures are extremely resistant to biodegradation, if indeed they are degraded at all.

The components of petroleum sulfonates with which one should be concerned, therefore, are the alkyl aryl sulfonates, both the linear (LAS) and the non-linear (ABS). This is particularly true as alkyl aryl sulfonates themselves are used to some extent in EOR.

The non-linear ABS compounds are typified by tetrapropylene alkyl benzene sulfonate, a relatively inexpensive material manufactured by petrochemical companies, and widely used in synthetic detergents, replacing traditional soaps, until about 1966. According to Cain (1981), by the early 1960's the environmental concentration of alkyl benzene sulfonates (probably mostly the tetrapropylene compound) was in the 1–10 mg/liter range, enough to produce much foaming and to be very troublesome in sewage treatment plants because of its very low degree of biodegradability. The petrochemical industry took the lead in finding a solution to this environmental problem, and by the late 1960's detergents containing alkyl benzene sulfonates with a linear alkyl group were introduced. Inasmuch as these linear (LAS) compounds are much more biodegradable than the tetrapropylene ABS, the excess foaming problem, with the consequent interference with the efficient operation of sewage treatment plants, was solved.

General formulas for the two types of compounds are as follows:

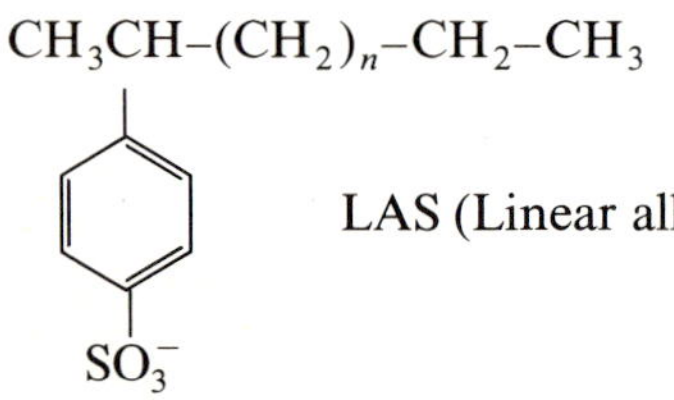

LAS (Linear alkyl benzene sulfonate)

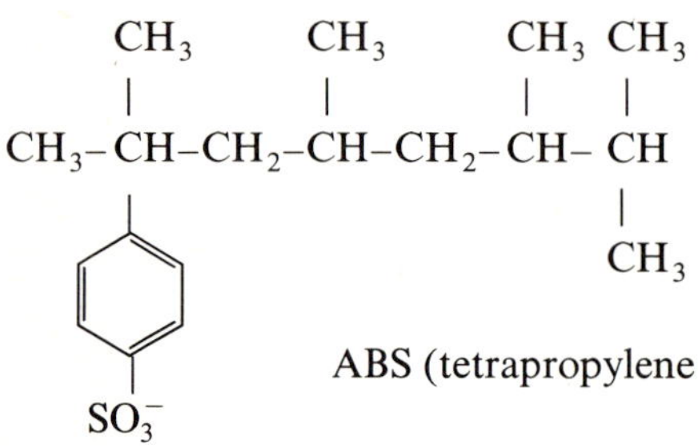

ABS (tetrapropylene benzene sulfonate)

In the case of LAS, the length of the carbon chain may vary, and the phenyl sulfonic acid group may be linked, in the *para* position, to any one of several carbon atoms in the linear R-group. Most LAS actually used in detergents probably contain a mixture of isomers.

Although it has been demonstrated on a large scale that tetrapropylene benzene sulfonate is resistant to biodegradation (recalcitrant would be an appropriate word), certain types of branching of the alkyl chain permit biodegradation to occur. If the structure of the branching permits β-oxidation of the alkyl group, biodegradation can take place (Swisher, 1970).

Cain (1981) provides a rather detailed discussion of possible mechanisms by which branched hydrocarbon chains may be degraded. It has been shown that certain branched ABS do undergo degradation (Huddleston and Allred, 1964; Swisher, 1970), although they possess structural features which preclude β-oxidation. Such features include substitution on the β-carbon, *gem*-dimethyl branching anywhere along the chain, or a quaternary branched carbon at the end of the alkyl chain. One possible mechanism for degradation of such compound is by α-oxidation in which one carbon is removed from the alkyl chain. The α-oxidation alone cannot completely degrade most branched chains. Certain types of branched chains cannot be broken down by either α-oxidation, or combined β- and α-oxidation (Cain, 1981). Two examples are:

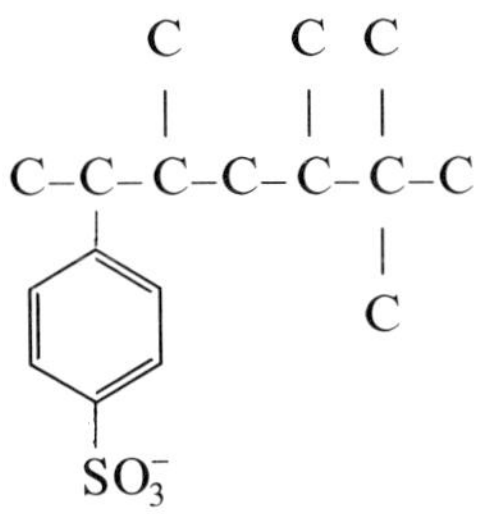

and:

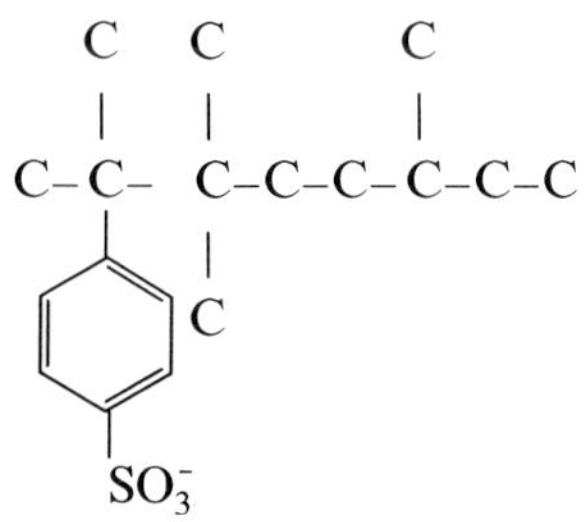

Certain natural products which are readily biodegradable have similar structures. Cain and his students (Willetts, 1973; Cain, 1981) have hypothesized that the catabolic mechanisms, by which dimethyl- and subterminal-substituted alkyl benzene sulfonates are degraded, parallel those already known for analogous natural products (for example, valine) and were perhaps catalyzed by the same enzymes. In particular a key enzyme would be 3-hydroxybutyryl-CoA hydrolase.

Cain (1981) further suggested that it is conceivable that certain microorganisms, under selection pressure to degrade ABS, could have evolved all the mechanisms necessary for efficient degradation of tetrapropylene benzene sulfonate, but this seems extremely unlikely in view of the failure of the extensive field testing to which this compound has already been subjected.

In contrast, it is generally agreed that LAS are readily biodegraded, and practical experience in treatment plants and water bodies confirm their relatively easy biodegradability (Nielsen and Huddleston, 1981).

Willetts and Cain (1970) have shown with pure culture studies that certain bacteria and fungi can completely degrade the linear alkylbenzene sulfonate molecule.

Regarding the mechanism of degradation, the alkyl chain is attacked first, by a process analogous to the oxidation of straight-chain-saturated aliphatic hydrocarbons. This proceeds by oxidation of the terminal-CH_3 to a COOH (ω-oxidation), followed by the stepwise removal of 2-C moieties by a process of β-oxidation. Degradation of the side chain is accompanied by a loss of methylene blue active substances (MBAS) (Longwell and Maniece, 1955), a test based upon surface activity of the compound. Loss of MBAS has been referred to as primary biodegradation. Ultimate or complete biodegradation would be conversion to cell material, CO_2, H_2O, and sulfate under aerobic conditions.

After alkyl group oxidation, the most common mode of attack involves desulfonation (this is indicated by release of inorganic sulfite to the medium, or by loss of UV absorption at 225 nm), followed by ring cleavage. However, at least six other routes of biodegradation have been reported (Cain et al., 1972). Nielsen and Huddleston (1981) stated that, inasmuch as such a variety of LAS metabolic capabilities is available in mixed microbial populations, the precise sequence of events during biodegradation of LAS under actual field conditions remains a challenging unknown.

It was recognized in early work that the biodegradation pathway for LAS consisted of many steps, and that the loss of surface activity occurred early in the biochemical sequence (Nielsen and Huddleston, 1981). Disappearance of surface activity coincided with the loss of reactivity to methylene blue (Swisher, 1973). This loss was later termed "primary biodegradation".

In recent years there has been some controversy regarding the complete or ultimate biodegradability of LAS. Results from two European laboratories indicated that *p-n*-alkylbenzene sulfonates are resistant to microbial attack (except for primary biodegradation), and that this resistance is mainly due to the sulfonic substituent (Leidner et al., 1980; Pitter and Fuka, 1979). The authors of these papers felt that, since earlier work on biodegradation of LAS was carried out using well-adapted pure cultures, the results might not reflect the behavior of the assayed compounds in natural aquatic environments.

Leidner et al. (1980) demonstrated that compounds with the same structure as the LAS, except for replacement of the sulfonic acid group by a carboxyl group, were readily degraded. They also found that the rates of breakdown of LAS with five or fewer carbon atoms between the sulfo-phenyl group and the terminal alkyl methyl group were markedly reduced below that of LAS with 12 carbon atoms, thus confirming the observations of Huddleston and Allred (1963). For example, toluene-*p*-sulfonate, *p*-1-propylbenzene-, *p*-1-butylbenzene-, and *p*-2-butylbenzene sulfonates resisted microbial attack under the conditions stated.

Pitter and Fuka (1979) stated that the ultimate biodegradability of linear alkylbenzene sulfonates does not occur even in long-term activated-sludge processes as determined by COD or total organic carbon. They further stated that the wastewaters usually contain some remaining biodegradation-resistant substances which no longer have characteristics of anionic active surfactants but are still aromatic in nature. These observations are in accord with many others to the effect that the alkyl chain (which is one factor in giving the molecule its surface activity) is degraded first. The persistence of the aromatic sulfonated portion of the molecule, however, is not in agreement with the findings reported by Nielsen and Huddleston (1981). Using ^{14}C-labelled (both in ring and in alkyl group) substrates, they indicated that the entire molecule, including the aromatic ring, was completely mineralized.

The sources of disagreement between the European and American groups include any one or more of the following: (1) differences in substrates used, (2) type of inoculum, or (3) approach and techniques used.

Leidner et al. (1980) compared a series of alkyl benzene sulfonates with the corresponding carboxylated compound for resistance to biodegradation. Their substrates only contained alkyl chains of 1, 3 and 4 carbon atoms; compounds which Huddleston and Allred (1963) had previously shown to be resistant to biodegradation. The main conclusion they reached was that the sulfonic acid group on an aromatic ring confers a degree of resistance to biodegradation that the carboxyl group does not. On the basis of their results, this appears to be a valid conclusion.

The nature of the inoculum would certainly seem to be an important factor in determining the rate of biodegradation of a given compound, if not its ultimate biodegradability. Several authors have pointed out the importance of using an "acclimated" or an "adapted" population in studies on the biodegradability of *n*-alkylbenzene sulfonates (Hollis and Phillips, 1976; Leidner et al., 1980; Grula and Grula, 1983). Most organisms capable of degrading these compounds in pure culture have been isolated from sewage (Willetts and Cain, 1972b; Bird and Cain, 1974), in which it would be expected that an enrichment of species of microorganisms possessing the necessary complement of enzymes for the degradation of these compounds would have occurred. Hollis and Phillips (1976) showed that acclimitized * (*sic*) inocula showed a definite increase in the rate of biodegradation of LAS (compound used was identified only as: LAS Reference Standard obtained from U.S. Environmental Protection Agency, NERC, Cincinnati, Ohio). Larson and Payne (1981) observed that the rate of degradation of LAS was reduced when using river water and sediment samples collected in the stream from above a sewage treatment plant, as compared to those samples collected downstream from the plant. However, the ^{14}C label used in the substrate (a C_{12} LAS, and a $C_{11.6}$ LAS, with a varied phenyl distribution) was bound to the particulates in the sediment two- to threefold more strongly in samples taken below the sewage outlet as compared to samples taken above. This reflects an increase in biomass (34-fold) occurring in the river below the outlet.

Larsen and Payne (1981), using appearance of radioactive CO_2 from ^{14}C-labelled ring, showed that half-lives for the LAS benzene ring were about 14 days in water samples from above a sewage outlet, and 1.4 days in samples from below the outlet. Corresponding figures for primary degradation of LAS ** caused by the two samples were not given. A period of acclimatization to ring cleavage, taking place in

* Fresh sewage inoculum incubated for 14 days at 25°C with LAS, and with air bubbled through mixture. During this period, LAS-degrading microorganisms presumably would be selected.

** Loss of surface activity; shown by decrease in methylene blue active substances (MBAS).

the sewage plant, was clearly indicated, whereas the necessity for acclimatization to primary biodegradation was not.

In their study, Nielsen and Huddleston (1981) used an acclimatized (10 mg/liter of unlabelled surfactant was added and mixture incubated until the maximum rate of LAS ring cleavage was achieved—4 days) inoculum from the Enid, Oklahoma, municipal sewage-treatment plant. This appears to be an adequate acclimatization treatment. From this paper, and that of Larson and Payne (1981), it is clear that acclimatization is required to select a population capable of ring cleavage. Hollis and Phillips (1976) only determined the concentration of methylene blue active substances. This test detects primary biodegradation–removal of the alkyl group on the benzene ring, but the systems of Hollis and Phillips also showed a requirement for acclimatization.

Finally, with regard to methods used, the European workers monitored biodegradation by measuring COD and total organic carbon, whereas in the U.S. laboratories, removal of the LAS molecule was followed by determination of $^{14}CO_2$ released from a radioactively labelled substrate.

One of the important factors in the acclimatization of natural populations to a capability for ultimate degradation of LAS appears to be the ability to desulfonate the ring. One reason for the increased recalcitrance that the sulfonic acid group imparts to a molecule, and for the rarity of naturally occurring sulfonates is the thermodynamic stability of the C–S bond of sulfonic acid compounds as compared with compounds with a sulfate ester (C–O–S) linkage. The final product of sulfonate metabolism is sulfate, with sulfite being a transient intermediate *. The mechanisms of desulfonation and of subsequent cleavage and oxidation of the ring are not well understood. For additional references one may consult Cain (1981).

It is known that arylsulfonate (both naphthalene and benzene sulfonates)-degrading bacteria rarely, if ever, grow on or oxidize the corresponding hydrocarbon. Both pathways are plasmid-encoded (Brilon, 1979).

Evolution of an arylsulfonate degrader from a degrader of the corresponding hydrocarbon has been shown to occur in the laboratory (Brilon, 1979). The conversion was evidently the result of an extremely rare event, because it required a 17-month exposure in a chemostat of *Pseudomonas testosteroni* (which initially could utilize naphthalene but not naphthalene sulfonates), to low concentrations of sulfonates in the naphthalene culture medium. The variant could utilize naphthalene sulfonates but no longer grew with naphthalene. Apparently, a new oxygenase had evolved which oxygenated naphthalene sulfonates to sulfonohydrodiols. The latter decomposed spontaneously to sulfite and to 1,2-dihydroxynaphthalene (Cain, 1981).

* Cain (1981) reported that in every alkylbenzene sulfonate-degrading bacterium isolated in his laboratory, for which it has been difficult to demonstrate sulfite release into the medium, there was found high activity of an inducible sulfite oxidase, so that sulfite would be oxidized as fast as it was formed.

The pathways of both aromatic hydrocarbon degradation and arylsulfonate degradation are plasmid-encoded (Sagoo, 1978). The plasmid encoding the pathway of arylsulfonate degradation (ASL) may be lost, after 12–16 generations in the absence of an alkylbenzene sulfonate (*P. arvilla*), or after a much longer time in *P. testosteroni* (Cain, 1981).

The nature of the populations of microorganisms capable of degrading arylsulfonates (LAS, ABS, or naphthalene sulfonates) is not well understood.

Several different genera of bacteria have previously been shown to be able to degrade alkylbenzene sulfonates. Using enrichment cultures in minimal salts medium with 0.05% (w/v) undecylbenzene-*p*-sulfonate, Willetts and Cain (1972a) isolated a species of *Bacillus* able to use the compound as a sole source of carbon and sulfur. The inoculum for the enrichments was obtained from a sewage treatment plant receiving household sewage and domestic detergent effluents. This organism appeared to use an entirely different mechanism for removing the sulfonic acid group from the molecule than the variant *Pseudomonas* strain isolated by Brilon (1979). The *Bacillus* species isolated by Willetts and Cain (1972a) was shown to possess a desulfonating enzyme, an inducible enzyme which removed the sulfonic acid group. Sulfite-cytochrome *c* oxidoreductase concentration was increased sixfold by growth on various alkylbenzene sulfonate homologues compared with growth on undecylbenzene as sole carbon source and either cysteine or methionine as the sole sulfur source. This is in contrast to the variant *Pseudomonas*, which oxidized the naphthalene sulfonate to the corresponding dihydrodiol, followed by spontaneous decomposition to sulfite and 1,2-dihydroxy naphthalene (Brilon, 1979).

Bird and Cain (1972) isolated a *Vibrio* species, capable of using dodecylbenzene sulfonate as sole carbon and energy source, which utilized only the side chain of the molecule. Enzyme characteristics of this organism were consistent with its production of C_2 units by β-oxidation of the side chain.

Horvath and Koft (1972) reported isolating an organism from an enrichment culture which was capable of cleaving the sulfonate moiety from alkylbenzene sulfonate and of using this sulfonate as its source of sulfur for growth. This organism, designated *Pseudomonas* HK-1, was evidently able to degrade the tetrapropylene side chain; tertiary butyl alcohol and isopropyl alcohol were produced. The strain could not utilize the ABS as a sole source of carbon and energy; growth on ABS alone did not occur. Presence of phenol in the medium greatly enhanced its utilization. This enhancement, together with the necessity of another carbon-energy source, suggested to the authors that degradation of ABS by *Pseudomonas* HK-1 actively occurred via cometabolic instead of metabolic reactions. To the knowledge of the writers, no bacterium has been reported to grow on ABS (non-linear) as a sole carbon-energy source. This is consistent with very low (if any) rate of degradation of ABS in wastewater treatment systems.

It is interesting to note that Horvath and Koft (1972) found that, with *Pseudomonas* HK-1, desulfonation appeared to involve the incorporation of molecular oxygen into the aromatic nucleus to yield catechol.

Another example of the utilization of tetrapropylene benzene sulfonate by what were, in all probability, cometabolic reactions is given in a paper by Dhawan and Mishra (1976). These authors used 50 ppm tetrapropylene sodium alkyl benzene sulfonate as a sole source of carbon in a mineral medium in enrichment cultures. Four isolates were obtained, two of which were identified as being *Bacillus*, and one each as *Pseudomonas* and *Serratia*. None of these bacteria, when isolated in pure culture, was able to grow in a medium with ABS as a sole carbon and energy source. Degradation and low growth (loss of methylene blue active substances) did occur if yeast extract was added, and with glucose plus yeast extract, growth and degradation were considerably higher than with only yeast extract as a supplement. The authors concluded that ABS could be degraded only by way of cometabolism.

In studying primary and total biodegradation of linear alkylbenzene sulfonates, Divo and Cardini (1980) used *Pseudomonas* MICRO-C and mixed soil microorganisms. The MICRO-C alone could degrade the side chain of certain LAS, but not others. Intermediates not degraded by MICRO-C could be completely broken down by mixed microbial cultures, apparently starting by an attack on the aromatic ring. It is not known how or when the sulfonate group was removed.

It is possible that fungi may be involved in degradation of alkylbenzene sulfonates in natural environments (Dimkov and Konstantinova, 1981). These authors observed that pure cultures of *Pseudomonas fluorescens* and *Aspergillus* species were able to degrade the alkyl chain of certain ABS (exact nature of compounds not known).

The presence of one of 10 selected aromatic or aliphatic compounds all tended to decrease the ability of these two strains to degrade, except in the case of β-keto adipic acid, which possessed an activating effect on the process.

An interesting type of interaction between microbial cultures growing on a mixture of 1-napthalene sulfonic acid and 2-naphthalene sulfonic acid was reported by Brilon et al. (1981).

The original source of the microorganisms was activated sludge from the Gottingen, F.R.G., sewage plant. Initially the population was naphthalene-degrading, but by continuous culture methods in the presence of increasing amounts of 2-naphthalene sulfonic acid (2NS) and decreasing amounts of naphthalene, a population was obtained which could utilize 2NS as a sole carbon source. (Their procedure was similar to that reported earlier by Brilon for isolation of a strain of *Pseudomonas testosteroni*, which utilized naphthalene sulfonic acids.)

The 2NS-utilizing continuous culture could be further adapted to the utilization of 1NS. If the 2NS were replaced by a mixture of 30% 2NS and 70% INS, both isomeric naphthalene sulfonates were completely removed from the culture fluid within four further months of continuous enrichment. Brilon et al. (1981) also observed that the mixture of 1NS and 2NS could be replaced by pure 1NS, and the population density did not change. No colonies, however, could be obtained on a mineral–1NS agar medium. If a mixture of 30% 2NS and 70% INS, plus succinate, was used in the mineral agar medium, colonies were obtained; actually 32 isolates were made. Of these, it was shown using high-pressure liquid chromatography that four showed some turnover of both 1NS and 2NS. One of these with the highest activity was chosen for further investigation.

When this isolate was placed on succinate mineral agar, three different strains were obtained, which were distinguished by their colony forms. None of these strains could grow singly on a mineral medium containing both 1NS and 2NS. However, growth was observed when the three strains were recombined. These strains were designated C_{22}, R_{22}, and T_{22}. A separate strain isolated from the 2NS (sole carbon source) continuous culture, designated A_3, was able to utilize both naphthalene and 2NS as sole carbon sources. Strains A_3 and C_{22} exhibited most of the phenotypic characters of the species *Pseudomonas testosteroni*. The strains R_{22} and T_{22} were not involved in the degradation of naphthalenesulfonic acid, and were designated by Brilon et al. (1981) as "escorting strains".

Both strains A_3 and C_{22} utilized 2NS as a sole source of carbon and energy. When strain C_{22} was grown in pure culture it required some vitamin B_{12}.

Sulfite accumulation in the medium was found with both organisms. All of the organic sulfur was finally converted to sulfate. Strain *Pseudomonas* C_{22}, when growing on pure 1NS, showed a considerably longer doubling time than it did on 2NS (26 hours, as compared to 7.5 hours).

The role of the escorting strains R_{22} and T_{22} is probably to supply certain growth factors for C_{22}. As indicated previously, none could grow singly on a mineral salts medium with naphthalenesulfonic acid as a sole carbon source. Optimum growth rates were found with a mixed culture of the three strains C_{22}, R_{22}, and T_{22} in the ratio 10 : 1 : 1 (as determined by plating). Some growth of C_{22} occurred on 70% 1NS/30% 2NS mineral salts medium when supplemented with yeast culture and vitamin B_{12}, but this growth was inferior to that occurring in mixed culture with the three strains. The C_{22} culture filtrates (from exponential culture growing with naphthalenesulfonates in the presence of B_{12}) would allow growth of R_{22} and T_{22} strains.

Pseudomonas sp. A_3, when subcultured on non-selective agar plates, showed an irreversible loss of degradative capacity for 2-naphthalenesulfonate, and also the ability to utilize naphthalene. The gene(s) for these abilities is (are) carried on a plasmid. Strains which have lost the plasmid can have it restored by conjugation with the wild type. The number of transconjugants per donor cell was found to be 0.6×10^{-6} to 1.0×10^{-6}.

One capability which is lost when the wild type is "cured" of its plasmid, apparently is the ability to oxidize naphthalene, since cured cells retain the ability to utilize salicylate, an oxidation product of naphthalene.

In spite of the excellent work by Brilon and his associates in characterizing microorganisms which degraded naphthalene sulfonic acid, the nature of the microorganisms found in the natural environment which are capable of degrading linear alkylbenzene sulfonates is not well understood. As in the case of *Pseudomonas* C_{22}, it may be that in most cases a consortium of species or at least two different microorganisms are required to bring about complete degradation of LAS molecules. Some authors (Willetts and Cain, 1972b; Bird and Cain, 1974) have isolated species of *Bacillus, Vibrio*, or *Alcaligenes* which could grow on certain LAS (usually dodecylbenzene sulfonate). Slater and Somerville (1979) quote Johanides and Hrsak (1976) as reporting that continuous flow enrichment culture with LAS as the limiting carbon-energy source produced a stable mixed culture which was dominated by species of *Pseudomonas* and *Alcaligenes*. None of the isolates from this enrichment culture was able to grow on LAS. Johanides and Hrsak (1976) are further quoted as suggesting that the degradation of the surfactant depended on the concerted activity of the complete community.

Cosurfactants, biocides, chelating agents and oxygen scavengers

Silvestro and Desmarais (1980) provided information on amounts used of various EOR chemicals in each of these classes. Probably other chemicals for these functions have been introduced since that time. Inorganic compounds are obviously not biodegradable in the usual sense. Many inorganic ions are toxic for microorganisms in high concentrations, even though they might be essential nutrients in low concentrations. Other inorganic materials, which could be used in EOR, are toxic even in low concentrations.

In regard to specific compounds, most of the alcohols used as cosurfactants are biodegradable, but the degree of biodegradability may vary considerably. Tertiary butyl alcohol is much less biodegradable than primary butyl alcohol (1-butanol), as indicated by the abundance of bacteria in soils able to use these compounds as sole carbon sources (Grula and Grula, 1983). Isobutyl alcohol is very readily degraded, whereas 2-butanol, with a secondary OH group, is much less readily degraded than either 1-butanol or 2-methyl-1-propanol (isobutyl alcohol) (Grula and Grula, 1983). Nevertheless, 2-butanol can be degraded. Grula and Grula (1983) isolated a *Geotrichum* species which grew well on 2-butanol as a sole carbon-energy source.

Not much information exists on biodegradability of biocides, chelating agents, or oxygen scavengers. Also, very little information is available on the rates of degradation in natural environments, types of microorganisms involved, or mechanisms of breakdown. A survey of the relative abundance of microorganisms capable of utilizing those compounds in various soils is given by Grula and Grula (1983).

REFERENCES

Alexander, M., 1981. Biodegradation of chemicals of environmental concern. *Science*, 212: 132.

Allred, R.C., Miles, T.A. and Fisher, H.B., 1954. Bacteriological techniques applicable to the control of sulfate-reducing bacteria in waterflooding systems. *Prod. Mon.*, December, pp. 31–32.

Bird, J.A. and Cain, R.B., 1972. Metabolism of linear alkylbenzene sulphonates by a *Vibrio* sp. *Biochem. J.*, 129: 46.

Bird, J.A. and Cain, R.B., 1974. Microbial degradation of alkylbenzene sulphonates. Metabolism of homologues of short alkyl-chain length by an *Alcaligenes* species. *Biochem. J.*, 140: 121–134.

Bourquin, A.W. and Pritchard, P.H. (Editors), 1979. *Proceedings of the Workshop: Microbial Degradation of Pollutants in Marine Environments, April 1978.* U.S. Environmental Protection Agency, Gulf Breeze, Fla.

Brilon, C., 1979. Ph.D. Thesis, University of Gottingen, Gottingen.

Brilon, C., Beckmann, W., Hellwig, M. and Knackmuss, H.J. 1981. Enrichment and isolation of naphthalenesulfonic acid-utilizing pseudomonas. *Appl. Environ. Microbiol.*, 42: 39–43.

Cadmus, M.C. and Slodki, M.E., 1984. Enzymic breakage of xanthan gum solution viscosity in the presence of salts. *Proc., Natl. Meet. Soc. Ind. Microbiol., Fort Collins, Color., August 1984.*

Cadmus, M.C., Jackson, L.K., Burton, K.A., Plattner, R.D. and Slodki, M.E., 1982. Biodegradation of xanthan gum by *Bacillus* sp. *Appl. Environ. Microbiol.*, 44: 5–11.

Cain, R.B., 1981. Microbial degradation of surfactants and "builder" compounds. In: T. Leisinger, A.M. Cook, R. Hutter and J. Muensch (Editors), *Microbial Degradation of Xenobiotics and Recalcitrant Compounds*. Academic Press, London, pp. 325–370.

Cain, R.B., Willetts, A.J. and Bird, J.A. 1972. Surfactant biodegradation: metabolism and enzymology. In: A.W. Walters and E.H. Hueck van der Plus (Editors), Applied Science Publishers, London, pp. 136–144.

Chu, J.P. and Kirsch, E.J., 1972. Metabolism of pentachloraphenol by an axenic bacterial culture. *Appl. Microbiol.*, 23: 1033–1035.

Dhawan, N. and Mishra, M.M., 1976. Biodegradation of tetrapropylene alkyl benzene sulphonate. *Zentralb. Bakteriol., Parasitenkd., Infektionskr. Hyg., Abt. 2*, 131: 208–211.

Dimkov, R. and Konstantinova, R., 1981. Biologisches Abbau von Alkylbenzolsulfonaten in Anwesenheit anderer organischer and anorganischer Verbindungen. *Zentralb. Bakteriol., Parasitenkd., Infektionskr. Hyg. Abt. 2*, 136: 161–167.

Divo, C. and Cardini, G., 1980. Primary and total biodegradation of linear alkylbenzenesulfonates. *Tenside Deterg.*, 17: 30–36.

Dorm, E., Hellwig, M., Reinecke, W. and Knackmuss, H.-J., 1974. Isolation and characterization of a 3-chlorobenzoate degrading strain. *Arch. Microbiol.*, 99: 61–70.

Edgehill, R.U. and Finn, R.K., 1983. Microbial treatment of soil to remove pentachlorophenol. *Appl. Environ. Microbiol.*, 45: 1122–1125.

Focht, D.D. and Williams, F.D., 1970. The degradation of *p*-toluene sulfonate by a *Pseudomonas. Can. J. Microbiol.*, 16: 309–316

Foshee, W.C., Jennings, R.R. and West, T.J., 1976. Preparation and testing of partially hydrolyzed polyacrylamide solutions. *Soc. Pet. Eng. of AIME, 51st Annu. Fall Tech. Conf., New Orleans, La., October 3–6*, SPE 6202.

Furukawa, K., Tomizuka, N. and Kamibayashi, A., 1983. Metabolic breakdown of Kanechlors (polychlorobiphenyls) and their products by *Acinetobacter* sp. *Appl. Environ. Microbiol.*, 46: 140–145.

Geffen, T.M., 1976. Present technology for oil recovery. *Proc., 1976 Eng. Found. Conf., The Role of Microorganisms in the Recovery of Oil, Easton, Md.*

Grula, M.M. and Huang, M., 1981. Interactions of polyacrylamides with certain soil pseudomonads. *Dev. Ind. Microbiol.*, 22: 451–457.

Grula, M.M. and Grula, E.A., 1983. *Biodegradation of Materials Used in Enhanced Oil Recovery*. Final Report, Contract DE-AT19-78BC-30201, U.S. Department of Energy.

Grula, M.M. and Sewell, G.W., 1983. Microbial interactions with polyacrylamide polymers. In: E.C. Donaldson and J.B. Clark (Editors), *Proceedings, 1982 International Conference on Microbial Enhancement of Oil Recovery*. U.S. Department of Energy, NTIS, Springfield, Va.

Hollis, C.H. and Phillips, W.K., 1976. Thermal effects on biodegradation of linear alkylbenzene sulfonates. *Dev. Ind. Microbiol.*, 17: 345–350.

Holzworth, G., 1984. Xanthan and scleroglucan: structure and use in oil recovery. *Proc., Natl. Meet. Soc. Ind. Microbiol., Fort Collins, Color., August 1984.*

Horvath, R.S. and Koft, B.W., 1972. Degradation of alkyl benzene sulfonates by *Pseudomonas* species. *Appl. Microbiol.*, 23: 407–414.

Howard, P.H., Saxena, J., Durkin, P.R. and Ou, L.F., 1975. *Evaluation of Available Techniques for Determining Persistence and Routes of Degradation of Chemical Substances in the Environment*. Office of Toxic Substances, U.S. Environmental Protection Agency, Washington, D.C.

Hršak, D., Bošnjak, M. and Johanides, V., 1981. Kinetics of linear alkylbenzene sulphonate and secondary alkane sulphonate biodegradation. *Tenside Deterg.*, 18: 137–140.

Huddleston, R.L. and Allred, R.C., 1963. Microbial oxidation of sulfonated alkylbenzenes. *Dev. Ind. Microbiol.*, 4: 24–38.

Huddleston, R.L. and Allred, R.C., 1964. The effect of structure on biodegradation of nonionic surfactants. In: F. Asinger (Editor), *Chemistry, Physics, and Applications of Surface Active Substances, Vol. III*. Gordon and Breach, London, pp. 871–880.

Jansson, P.E., Kenne, L. and Lindberg, B., 1975. Structure of the extracellular polysaccharide from *Xanthonomas campestris*. *Carbohydr. Res.*, 45: 275–282.

Johanides, V. and Hršak, D., 1976. Changes in mixed bacterial cultures during linear alkylbenzene-sulfonate (LAS) biodegradation. In: H. Dellweg (Editor), *Fifth International Symposium on Fermentation*. Westkreuz, Berlin, p. 426.

Kang, K.S. and Cottrell, I.W., 1979. Polysaccharides. In: H.J. Peppler and D. Perlman (Editors), *Microbial Technology, Vol. 1*. Academic Press, London, 2nd ed., pp. 417–481.

Klausmeier, R.E. and Andrews, C.C., 1981. Plastics. In: A.H. Rose (Editor), *Microbial Biodeterioration. Economic Microbiology, Vol. 6*. Academic Press, London, pp. 431–474.

Knackmuss, H.J. and Hellwig, M., 1978. Utilization and cooxidation of chlorinated phenols by *Pseudomonas* sp. B 13. *Arch. Microbiol.*, 117: 1–7.

Kravetz, L., Chung, H. and Rapean, J.E., 1982. Ultimate biodegradation studies of α-olefin sulfonates. *J. Am. Oil. Chem. Soc.*, 59: 206–210.

Larson, R.J. and Payne, A.G., 1981. Fate of the benzene ring of linear alkylbenzene sulfonate in natural waters. *Appl. Environ. Microbiol.*, 41: 621–627.

Leidner, H., Gloor, R., Wuest, D. and Wuhrmann, K., 1980. The influence of the sulphonic group on the biodegradability of *n*-alkylbenzene sulphonates. *Xenobiotica*, 10: 47–56.

Leisinger, T., Cook, A.M., Huter, R. and Nuensch, J. (Editors), 1981. *Microbial Degradation of Xenobiotics and Recalcitrant Compounds. FEMS Symposium, 12*. Academic Press, London.

Longwell, J. and Maniece, W.D., 1955. Determination of anionic detergents in sewage, sewage effluents, and river water. *Analyst (London)*, 80: 167–171.

Maerker, J.N., 1975. Shear degradation of partially hydrolyzed polyacrylamide solutions. *Soc. Pet. Eng. J.*, 15: 311–312.

Maki, A.W., Dickson, K.L. and Cairns, J., Jr. (Editors), 1980. *Biotransformation and Fate of Chemicals in the Aquatic Environment*. American Society for Microbiology, Washington, D.C.

Meyers, J.D. and Huddleston, R.L., 1979. Treatment of Oily Refinery Wastes by Landfarming. *34th Ann. Purdue Ind. Waste Conf., West Lafayette, Ind., May 1979* (presentation).

Nielsen, A.M. and Huddleston, R.L., 1981. Ultimate biodegradation of linear alkylbenzene sulfonate alkyl and ring carbon. *Dev. Ind. Microbiol.*, 22: 415–424.

Pitter, P. and Fuka, T., 1979. The problem of ultimate biodegradability of linear alkylbenzene sulphonates. *Tenside Deterg.*, 16: 298–302.

Postgate, J.R., 1979. *The Sulphate-Reducing Bacteria*. Cambridge University Press, Cambridge.

Ratledge, C., 1978. Degradation of aliphatic hydrocarbons. In: R.J. Watkinson (Editor), *Developments in Biodegradation of Hydrocarbons, Vol 1*. Applied Science Publishers, London, pp. 1–46.

Ripin, M.J., Noon, K.F. and Cook, T.M., 1971. Bacterial metabolism of aryl sulfonates. 1. Benzene sulfonate as growth substrate for *Pseudomonas testosteroni* H-8. *Appl. Microbiol.*, 21: 495–499.

Rose, A.H. (Editor), 1981. *Economic Microbiology, Vol. 6. Microbial Biodeterioration.* Academic Press, London.

Sagoo, G.S., 1978. M.Sc. Thesis, University of Kent, Canterbury.

Sandford, P.A., Pittsley, J.E., Knutson, C.A., Watson, P.R., Cadmus, M.C. and Jeanes, A., 1977. Extracellular microbial polysaccharides. In: P. Sandford and A. Laskin (Editors), *ACS Symp. Ser., 45*. American Chemical Society, Washington, D.C., pp. 192–210.

Schwein, U. and Schmidt, E., 1982. Improved degradation of monochlorophenols by a constructed strain. *Appl. Environ. Microbiol.*, 44: 33–39.

Silvestro, E. and Desmarais, A.-M., 1980. *Toxicity of Chemical Compounds Used for Enhanced Oil Recovery*. Final Technical Report, U.S. Department of Energy, BETC, Contract No. DE-RP19-79BC 10014.

Slater, J.H. and Somerville, H.J., 1979. Microbial aspects of waste treatment, with particular attention to the degradation of organic compounds. In: A.T. Bull, D.C. Ellwood, and C. Ratledge (Editors), *The Society for General Microbiology, Symposium 29*. Cambridge University Press, Cambridge, pp. 221–261.

Slodki, M.E. and Cadmus, M.C., 1982. Production, composition, and biodegradation of xanthan gum (polysaccharide B-1459). In: *Proceedings, International Conference on Microbial Enhancement of Oil Recovery, Shangri-La, Afton, Oklahoma, May 1982*. Conf. 8205140, U.S. Department of Energy, pp. 3–5.

Stanlake, G.J. and Finn, R.K., 1982. Isolation and characterization of a pentachlorophenol degrading bacterium. *Appl. Environ. Microbiol.*, 44: 1421–1427.

Swisher, R.D., 1969. Benzene ring biodegradation in quaternary alkylbenzene sulfonates. *Tenside Deterg.*, 6: 135–139.

Swisher, R.D., 1970. *Surfactant Biodegradation*. Marcel Dekker, New York, N.Y.

Swisher, R.D., 1973. Anionic surfactants and the environment: Discussion. *Progr. Water Technol.*, 3: 303–306.

Swisher, R.D., 1981. The problems of ultimate biodegradability of linear alkylbenzene sulfonates. *Tenside Deterg.*, 18: 57–63.

Swisher, R.D., 1982. Kinetics of sulfonate surfactant biodegradation. *Tenside Deterg.*, 19: 298–299.

Watanabe, I., 1973. Isolation of pentachlorophenol-decomposing bacteria from soil. *Plant Nutr. (Tokyo)*, 19: 109–116.

Willetts, A.J., 1973. Microbial aspects of the biodegradation of synthetic detergents. *Int. Biodeterior. Bull.*, 9: 3–10.

Willetts, A.J. and Cain, R.B., 1970. Microbial metabolism of alkylbenzene sulphonates. *Biochem. J.*, 120: 28p 29p.

Willetts, A.J. and Cain, R.B., 1972a. Microbial metabolism of alkylbenzene sulphonates. Bacterial metabolism of undecylbenzene-*p*-sulphonate and dodecylbenzene-*p*-sulphonate. *Biochem. J.*, 129: 389–402.

Willetts, A.J. and Cain, R.B., 1972b. Microbial metabolism of alkylbenzene sulphonates. Enzyme system of a *Bacillus* species responsible for β-oxidation of the alkyl side chain of alkyl-benzene sulphonates. *Antonie van Leeuwenhoek J. Microbiol. Serol.*, 38: 543–555.

Chapter 11

POTENTIAL HEALTH HAZARD OF BACTERIA TO BE USED IN MICROBIAL ENHANCED OIL RECOVERY

E.A. GRULA *, HUGH H. RUSSELL and MARY M. GRULA

INTRODUCTION

Because practically all life forms may be infected by one or more kinds of microorganisms, resulting in disease of the host, the use of bacteria in MEOR operations necessitates a consideration of possible untoward effects against man and other living creatures. It should be emphasized that there are more "good" or non-pathogenic bacteria than "bad" or pathogenic types. Because large numbers of bacteria, however, are going to be placed underground and possibly come into direct contact with oilfield workers who know little about them other than the fact that they are loosely referred to as "germs", it is necessary to examine closely possible hazards that might be associated with the routine use of bacteria in oilfield endeavors relating to EOR.

HEALTH HAZARD

Depending on the organism, bacteria can cause illness in human beings in several ways. Some bacteria must be continually present in the body as live entities to cause illness either directly or because of toxins they produce *in vivo*. Others produce their toxins outside the body and these compounds, when ingested, are responsible for the sickness. Needless to say, everyone knows that some illnesses are, to varying extents, merely debilitative, whereas others may be fatal.

Bacteria can enter the body by way of bites, cuts or abrasions, deep wounds, or via the sweat glands, ingestion or inhalation. When they are present throughout the blood stream, with or without portions of their cellular remains (for example, the lipopolysaccharide of some Gram-negative bacteria), a dangerous and septicemic condition exists. Rather than being generally distributed via the blood or lymph system, some bacteria remain localized in a particular organ or region of the body whereas others have the ability to spread from a localized area by penetration of the tissue cementing substance (hyaluronidase enzyme activity). Still others move from

* Deceased September 1, 1983

cell to cell by directly infecting adjacent cells which they first kill by elaboration of various cytotoxic chemicals. For a general reference on toxins (both exotoxins and endotoxins, portals of entry, and invasiveness of microbes), one should consult Davis et al. (1980).

In addition to infections, produced by active invasion of the body, bacterial preparations, including dried types such as lyophilized cells can, particularly in susceptible individuals, lead to varying degrees of allergic responses after inhalation. This type of response is different from a bacterial infection; preparations causing allergies may be of either dead or living cells. Wet bacterial preparations should not be a serious problem in producing this type of allergy, because it is highly unlikely that such preparations would be inhaled; formation of an aerosol from the preparation would be required in order for inhalation to occur. It is very improbable that an aerosol would be produced under oilfield conditions. A discussion of bacterial preparations as allergens may be found in Davis et al. (1980, chapter 22).

A point that should be made or perhaps even stressed is that many oilfield ponds and waters used in waterflooding processes contain various kinds and numbers (some very high) of microorganisms. Also, workers involved in enhanced recovery methodologies involving polymers are constantly exposed to many bacteria in waters being used. One cannot, therefore, take the position that MEOR technologies introduce a completely new element (i.e., bacteria) into the existing work place and, therefore, will require significant adjustments in oilfield worker behavior.

Bacteria that have been thought to be useful in MEOR, or to be present in wells as part of the indigenous microflora include *Desulfovibrio* spp., the methanogens, *Bacillus* spp., miscellaneous Gram-negative rods and members of the genus *Clostridium*.

The writers do not know of any deleterious infections of humans or animals caused by the seemingly ubiquitously distributed sulfate-reducers or methanogens. In addition, there are no current serious proposals for their extensive use in MEOR.

To the knowledge of the writers all pathogenic members of the genus *Bacillus* (*B. anthracis* and *B. cereus*) are aerobic; thus mitigating against their use in downwell MEOR operations. There is no evidence that facultative anaerobes belonging to this genus, which might be useful in MEOR, are pathogenic.

Because of their ability to incorporate plasmids of various kinds, including those involved in biodegradation of hydrocarbons, some members of the genus *Pseudomonas* are candidates for MEOR. Since most of their proposed use is based on "partial" utilization of reservoir hydrocarbons (usually in order to reduce viscosity), economic factors that relate directly to loss of the product being sought is a serious obstacle to their consideration and use. Also, the most expeditious and by far the most frequent method of microbial degradation of hydrocarbons (ring or aliphatic chains) necessitates the use of oxygen-inserting enzymes (mono- and dioxygenases) (Watkinson, 1978). Thus, the anaerobic conditions of a reservoir constitute a serious impediment to use of pseudomonads, particularly where continued underground movement of the bacteria will be necessary as in a line-drive type recovery process.

Gas production (particularly carbon dioxide because its miscibility in oil causes reduction in oil viscosity) appears to be a very desirable and important attribute of any bacterial culture to be selected for use in MEOR. Of all Gram-negative bacteria, *Enterobacter aerogenes* (formerly *Aerobacter aerogenes*) appears to be a logical choice because ratios of CO_2 to H_2 produced by this organism are very high (at least 2 to 1). To the knowledge of the writers, *E. aerogenes*, which can exist as a free-living saprophyte does not cause any disease; however, considering the ability of the enteric bacteria (*E. aerogenes* can generally be found inhabiting the human intestinal tract about 50% of the time) to exchange genetic information, vigilance should be exercised in the event urinary tract infections or intestinal disturbances begin to appear in workers handling such organisms.

Some investigators (Jenneman et al., 1983) believe that certain Gram-negative isolates can be used to seed highly permeable regions of a reservoir (oil-containing sands, channels, etc.). During growth, it is proposed that these organisms will elaborate highly viscous exocellular polymers thus plugging such areas and resulting in more sweep efficiency and oil recovery using waterflooding. To the knowledge of the writers, all of the organisms proposed for such MEOR operations are non-pathogenic Gram-negative types.

The microorganisms most often proposed for (or, indeed, actively utilized in) MEOR operations at the present time are members of the genus *Clostridium*. This emphasis exists because several members of this genus form large amounts of CO_2 and other desirable metabolic end-products (acids, solvents, emulsifiers; see Chapter 10), grow well in "simple" molasses media, are anaerobic, and form hardy spores. Some members of this genus are also responsible for several manifestations of illness in man. Included are severe food intoxications (*C. botulinum*), gas gangrenous infections (e.g., *C. perfringens*), and tetanus (*C. tetani*). These clostridia are not parasitic, nor do they invade the body in the way that most pathogenic bacteria do; they produce disease by synthesizing an exotoxin which then gains access to the body via ingestion (botulinum toxin), from a very localized puncture wound (tetanus toxin), or via invasion of necrotic tissue (*C. perfringens*) (Davis et al., 1980). Certain strains of *C. perfringens* produce a foodborne illness; this is the result of certain specific exotoxins (enterotoxins) elaborated by the strain. The toxin is synthesized by sporulating cells, and the peak for toxin synthesis is reached just before lysis of the sporangium. The toxin is released along with the spores. The enterotoxin is quite heat sensitive and is destroyed by normal cooking procedures (Riemann and Bryan, 1979). Ordinarily, to produce the illness, live cells must be ingested with the food. They sporulate freely in the gastrointestinal tract. In a somewhat similar fashion, certain strains of *C. botulinum* may gain access to the digestive tract of babies, and produce the syndrome known as infant botulism. In general, however, clostridia are not invasive, and produce disease only through the secretion of exotoxins.

The genus *Clostridium* includes many organisms that are non-pathogenic and which have been used for many years in very large industrial processes (*C. acetobutylicum* is a prime example) without untoward effects to the workers involved. It appears, therefore, that the judicious laboratory selection of non-pathogenic clostridia possessing the ability to produce the desired metabolic end-products will most likely, at least in the near future, be the most significant factor in the prevention of accidentally produced bacterial disease in oilfield workers. Non-pathogenic clostridia will probably be the organisms most often used in "huff and puff" (and probably line-drive) -type MEOR technology.

Certain rules pertaining to extensive use of bacteria in the oilfield appear to be desirable. First and foremost, a competent microbiologist should be in charge of the operations, particularly those aspects relating to handling of the organisms. An important duty would be the instruction of oilfield workers in common sense behavior concerning the handling of and contact with bacteria. For example, cuts and abrasions should never be permitted to come into contact with the bacteria. If this does happen, soap, water, and disinfectants should be immediately available and their use encouraged. Also, inhalation of the microorganisms should always be kept to an absolute minimum by use of face masks and by allowing the wind to blow the organisms away from the operator. The risk of inhalation seems minimal since the formation of an aerosol from the spore suspensions prepared for injections would be necessary. On-site surveillance of worker activity should be an integral part of the operation. Also, to help the microbiologist, the number of workers directly involved in the handling of bacterial preparations should be limited and he/she should have complete authority in this area of the total endeavor.

Without question, the microbiologist in charge of the operation should have been involved in the laboratory screening procedures in which it was determined that the cultures being used are non-pathogenic. Based on information given in the 8th edition of Bergey's manual (Buchanan and Gibbons, 1974), five simple biochemical tests can be used to screen out clostridia which may be pathogenic.

They are: (1) proteolytic activity in meat broth, (2) hydrolysis of casein, (3) beta haemolysis of red blood cells (blood agar plates), (4) production of hydrogen sulfide, and (5) fermentation of sucrose. Although a clostridial isolate may possess one or more of these characteristics, it is to be emphasized that *all* need to be present to allow strong presumptive evidence of pathogenicity. Such cultures should be immediately discarded. Possession of any three might warrant further testing to rule out pathogenicity. Such confirmatory final testing, if needed, should involve the proper use of animals.

Considering the large investments involved as well as the health of people, it is very desirable that enough of both time and money are spent in laboratory isolation and screening procedures of bacterial strains before any field application of the strains in MEOR operations. Certainly, the use of raw sewage should never be permitted because many opportunistic pathogens are normal inhabitants of the human gastrointestinal tract (e.g., *C. perfringens*). A significant number of apparently healthy individuals are carriers of *Salmonella* (Riemann and Bryan, 1979). One species, *Salmonella typhi*, is the cause of typhoid fever.

Although highly impermeable layers of shale or other caprocks cover most oil-bearing strata, it is, nevertheless, the responsibility of microbiologists to, at all costs, avoid contamination of underground aquifers or surface waters by seepage resulting in the underground movement of bacteria. In short, all microbiological "insults" to the environment should be avoided. It is necessary to view all activities in a critical manner and to give full consideration to the environment and its inhabitants who might be adversely affected by these activities.

The knowledge gained from a limited field experience is of some value. Bacteria are transported to the site in plastic containers, or in 55-gallon drums. At the wellsite, work is performed as carefully as humanly possible. Bacteria are either poured into the casing from the plastic containers or a centrifugal pump (Deaps model). A bull plug with a 1-inch hose connection is used to pump the inoculum from the 55-gallon drum. The microbiologists should arrive before the trucks and other workers, so that the bacteria are ready to be injected immediately.

In summary, provided that certain precautions are taken there appears to be no valid or serious objections to the use of *selected* strains of bacteria in MEOR operations. The bacteria should be screened for possible pathogenicity and the microbiologist in charge of the operations should exert rigorous control and dispense the technical information required to maintain the health and safety of oilfield workers and the environment.

REFERENCES

Buchanan, R.E. and Gibbons, N.E. (Co-Editors), 1974. *Bergey's Manual of Determinative Bacteriology*. Williams and Wilkens, Baltimore, Md., 8th ed.

Davis, B.D., Dulbecco, R., Eisin, H.N. and Ginsberg, H.S., 1980. *Microbiology* Harper and Row, Hagerstown, Md., 3rd ed.

Jenneman, G.E., Knapp, R.M., Menzie, D.E., McInerney, M.J., Revus, D.E., Clark, J.B. and Munnecke, D.M., 1983. Transport phenomena and plugging in Berea Sandstone using microorganisms. In: *Proceedings, 1982 International Conference on Microbial Enhancement of Oil Recovery*. U.S. Department of Energy, NTIS, Springfield, Va., p. 71.

Riemann, H. and Bryan, F.L. (Editors), 1979. *Flood-Borne Infections and Intoxications*. Academic Press, New York, N.Y., 2nd ed.

Watkinson, R.J. (Editor), 1978. *Developments in Biodegradation of Hydrocarbons, Vol. 1*. Applied Science Publishers, London.

REFERENCES INDEX *

---

* Assistance in preparation of the indices by Baljit Singh is greatly appreciated by the editors.

SUBJECT INDEX